BASEMENT TECTONICS 11

Proceedings of the International Conferences on Basement Tectonics

VOLUME 5

The titles published in this series are listed at the end of this volume.

BASEMENT TECTONICS 11

Europe and Other Regions

Proceedings of the Eleventh International Conference
on Basement Tectonics, held in Potsdam,
Germany, July 1994

Edited by

O. ONCKEN
and
C. JANSSEN
GeoForschungsZentrum Potsdam, Germany

SPRINGER-SCIENCE+BUSINESS MEDIA, B.V.

A C.I.P. Catalogue record for this book is available from the Library of Congress.

ISBN 978-94-010-7216-8 ISBN 978-94-009-1598-5 (eBook)
DOI 10.1007/978-94-009-1598-5

Printed on acid-free paper

Originally published by Kluwer Academic Publishers in 1996
Softcover reprint of the hardcover 1st edition 1996

CONTENTS

CONFERENCE COMMITTEES

SCIENTIFIC COMMITTEE

A. Zelazniewicz, Institut Nauk Geologicznych PAN, ul. Podwale 75, Pl 50-449 Wroclaw, Polen.

Z. Pertold, Department Geology of Mineral Deposits, Charles University, 12843 Praha, Czech Republic.

H. Kern, Mineralogisches Institut , Universität Kiel, Olshausenstraße 40-60, 24118 Kiel.

St. Mueller, Institut für Geophysik, ETH Zürich, Hönggerberg, CH-8093 Zürich.

ORGANIZING COMMITTEE

O. Oncken, GeoForschungsZentrum Potsdam, Telegrafenberg C 2, 14473 Potsdam, Germany.

P. Bankwitz, GeoForschungsZentrum Potsdam, Telegrafenberg A 3, 14473 Potsdam, Germany.

G. Dresen, GeoForschungsZentrum Potsdam, Telegrafenberg A 26, 14473 Potsdam, Germany.

V. Haak, GeoForschungsZentrum Potsdam, Telegrafenberg A 46, 14473 Potsdam, Germany.

SPONSORS

The 11th international conference on 'Basement Tectonics' held in Potsdam, from July 25th to July 29th 1994, was sponsored by:

Deutsche Forschungsgemeinschaft DFG

GeoForschungsZentrum Potsdam GFZ

Ministry for Science, Research and Culture of the state of Brandenburg MWFK

University of Potsdam

FOREWORD

The Eleventh International Conference on Basement Tectonics was held at the GeoForschungsZentrum Potsdam in Germany 25-29 July 1994. It was the first time that a meeting of this series took place in the centre of Europe and the first such meeting in this area after the recent political changes in eastern Europe. Consequently, the main theme of the meeting focussed on the structure and history of the entire European continent. Further themes were grouped around topics of current interest. Keynote lectures on these topics were given by A. Berthelsen, St. Mueller, A. Green, and D. Fountain. The technical sessions were arranged with support of the Scientific Committee on the following topics:

1. Continental scale features of basement rocks of phanerozoic cratons - with emphasis on Europe
2. Extrapolation and correlation of geological and geophysical data from basement rocks: an assessment
3. Mechanisms of basement exhumation in the evolution of orogenic belts
4. Structure and intraplate deformation of the North-American craton

Papers submitted to this volume have however all been collected in one part since all of them deal with different aspects of deformation of the continental basement. A second part contains a complete list of all papers and posters presented at the meeting.

Field trips to study some of the above mentioned aspects in the Harz mountains, the Granulite Massif, and the Erzgebirge were organized by P. Bankwitz, P. Frischbutter, J. Rötzler, K. Rötzler, and B. Mingram. Their assistance is gratefully acknowledged.

The meeting brought together 145 scientists from more than 25 countries, more than a third of them from east-European countries. Their participation as well as the entire meeting would never have been possible without the generous technical and financial support of several institutions: The GeoForschungsZentrum Potsdam as the organizing institution, the Ministry of Science, Research and Culture of the State of Brandenburg, the German Research Council (DFG), and the University of Potsdam. Their support is gratefully acknowledged.

Last not least the organizing committee wishes to thank all members from the GFZ which helped with the organisation, the registration desk, slide projection, transport etc. G. Wolff and M. Bach especially have helped to manage all technical challenges which arised before, during, and after the meeting. We thank all delegates which have helped with their attendance and contributions to make the meeting a success. Finally, we thank the Trustees of the International Basement Tectonics Incorporated for entrusting the Conference to Potsdam.

O. Oncken
Convener

PART 1

SELECTED PROCEEDINGS PAPERS

LATE-OROGENIC BURIAL AND EXHUMATION AT A MAJOR VARISCAN SUTURE ZONE

Constraints from Thermal Modeling

A. HENK
Institut für Geologie
Pleicherwall 1
D-97070 Würzburg
Germany

Abstract. A two-dimensional thermal-kinematic modeling approach constrained by geothermo-barometric and geochronological data was used to investigate the Carboniferous burial and exhumation history at a major Variscan suture zone in Central Europe. Modeling suggests that during Late Viséan to Early Namurian times convergence rates between two of the main tectonometamorphic units of the Variscan fold belt, Saxothuringian and Rhenohercynian, exceeded 30 mm/a and subsequently slowed down to 20 mm/a. Exhumation in the direct vicinity of the suture zone occurred temporarily at velocities of more than 3 mm/a. Such rapid decompression strongly suggests a significant contribution of extensional strain to the Late Variscan exhumation of the study area.

1. Introduction

Finite element techniques were applied to examine the crustal temperature field at a Variscan suture zone in Central Europe in relation to late-orogenic burial and exhumation (Fig. 1). This suture separates not only two of the main tectonostratigraphic units of the Variscan fold belt, the Rhenohercynian and Saxothuringian zones, but also forms the boundary between the external fold-and-thrust belt and the crystalline complexes of the orogen's interior. Its Late Variscan evolution was directly related to the final closure of the Rheno-

O. Oncken and C. Janssen (eds.), Basement Tectonics 11, 3–15.

hercynian or Lizard-Giessen-Harz Ocean (Franke & Oncken, 1990) situated between the Rhenohercynian and Saxothuringian zones. Closure of this small oceanic basin during the Upper Devonian changed the geodynamic position of the northern Saxothuringian zone, the so-called Mid-German Crystalline Rise (MGCR), from a passive to an active margin setting related to a SE-directed subduction zone (Franke & Oncken, 1990; Franke, 1992). Consumption of oceanic crust was rather short-lived and essentially completed at the end of the Devonian. Plate convergence continued throughout the Early Carboniferous and ultimately led to thrusting of the Saxothuringian zone onto the Rhenohercynian rocks to the north. As a result, rocks of the northern Saxothuringian and the southern Rhenohercynian zone went through very different tectonothermal histories during the final stage of the Variscan orogeny. Particularly during Late Viséan and Early Namurian times, exhumation of the MGCR was contemporaneous with the onset of convergence in the southern Rhenohercynian zone as well as continuous compression in the Saxothuringian basin

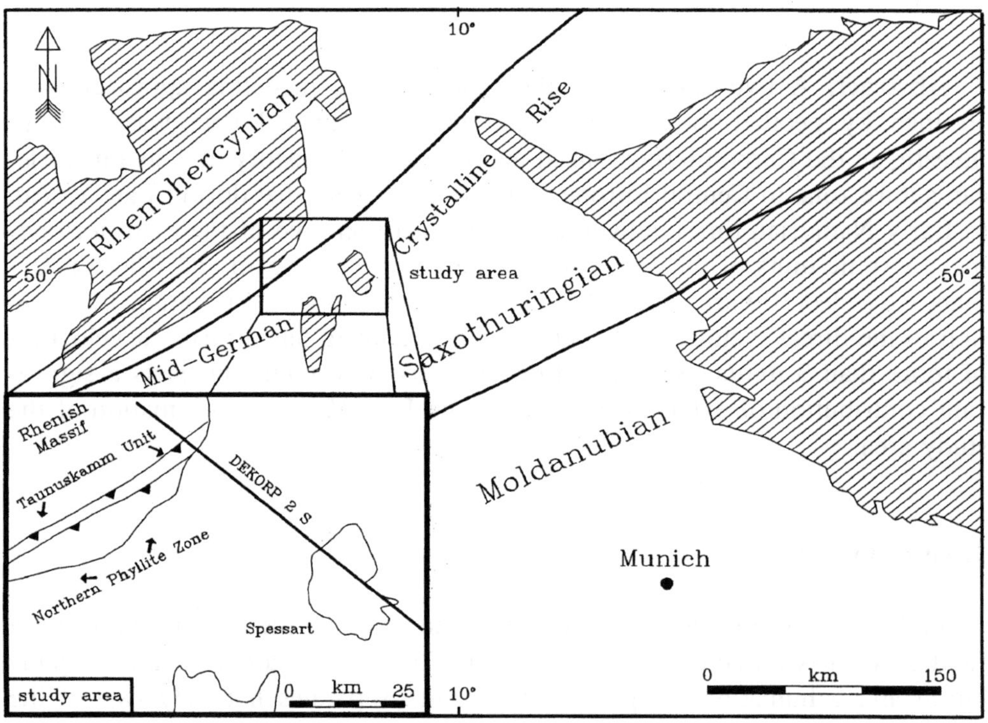

Figure 1. Simplified geological map showing the main tectonostratigraphic units of the Variscan fold belt in Central Europe and Variscan massifs of the study area (Taunuskamm Unit = TKU; Northern Phyllite Zone = NPZ).

farther south. The different evolutions are still preserved in variable pressure - temperature - time - (P-T-t-) paths. Deciphering of these P-T-t records by means of thermal modeling techniques can therefore provide not only quantitative estimates for late-orogenic exhumation velocities and thermal state of the Late Variscan crust, but also for convergence rates between Rhenohercynian and Saxothuringian.

2. Geological Data

The study concentrates on the Late Variscan tectonometamorphic evolution of three complexes located near a NW-SE oriented transect across the suture between the Rhenohercynian and Saxothuringian zones (Fig. 1). The crustal structure along this transsect was revealed by deep seismic line DEKORP 2 S. The interpretation by Behr & Heinrichs (1987) shows several major thrust zones with large-scale ramp structures, typically dipping at angles of 15 to 20° (Fig. 2).

The P-T-t data used to constrain thermal modeling are summarized in Table 1. The evolution of the MGCR or hanging wall with respect to the suture is represented by the Spessart crystalline complex. It consists mainly of gneisses and mica-schists which show a medium pressure type of amphibolite metamorphism (5 - 6.5 kbar, 570 - 620° C) as the main Variscan metamorphic event (Okrusch, 1990). The exhumation history is constrained by several K/Ar

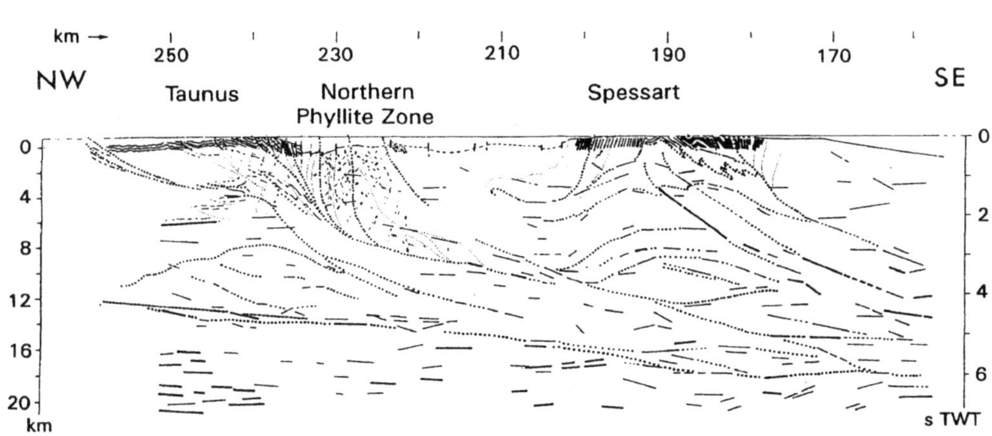

Figure 2. Line drawing of DEKORP 2S deep seismic reflection profile with interpretation of upper crustal structure and surface geology at the suture between the Saxothuringian and Rhenohercynian zones (simplified after Behr and Heinrichs, 1987). Vertical exaggeration: about 1.6 . See inset Fig. 1 for location.

data which give similar ages essentially independent of mineral type and hence closure temperature. The tectonometamorphic evolution of the suture's footwall is documented in the Northern Phyllite Zone (NPZ) and Taunuskamm Unit (TKU) in the southern Rhenohercynian zone. These narrow tectonic slices are separated by thrusts from the Rhenish Massif to the north. Sediments and volcanics of the NPZ experienced peak metamorphic pressures of

P (kbar)	T (°C)	t (Ma)	Description	Reference
Spessart				
5 - 6.5	570 - 620		main metamorphic event	e
530 ± 40		321 ± 4	K-Ar hornblende cooling age	d
350 ± 40		321 ± 4	K-Ar muscovite cooling age	d
300 ± 40		321 ± 3	K-Ar biotite cooling age	d
surface exposure		295 ± 5	detritus in Wetterau basin	c
Northern Phyllite Zone				
sedimentation		< ?330	inferred from stratigraphic and facial similarities to the TKU	a
5 - 6	300 - 330	325 ± 5	phengite barometry, static recrystallization of quartz, absence of biotite, K-Ar white mica formation age	a, b
5 ± 1	270 - 300	310 ± 5	phengite barometry, static recrystallization of quartz, K-Ar white mica formation age	a, b
3 ± 1	270 - 300	310 ± 5	phengite barometry, static recrystallization of quartz, K-Ar white mica formation age, no subsequent differential movement between NPZ and TKU	a, b
surface exposure		295 ± 5	detritus in Wetterau basin	c
Taunuskamm Unit				
sedimentation		< 330	Viséan 2 in Hintertaunus Unit	a
3 ± 1	270 - 300	320 ± 5	phengite barometry, static recrystallization of quartz, K-Ar white mica formation age	a, b
3 ± 1	270 - 300	310 ± 5	phengite barometry, static recrystallization of quartz, K-Ar white mica formation age	a, b
surface exposure		295 ± 5	detritus in Wetterau basin	c

Table 1. Available P-T-t-data (references: a = Anderle et al., 1990; b = Klügel & Oncken, 1994; c = Kowalczyk, 1983; d = Nasir et al., 1991; e = Okrusch, 1990).

5 - 6 kbar similar to the Spessart, but in contrast temperatures never exceeded 330° C as biotite is absent (Anderle et al., 1990). The burial and exhumation histories of both units are constrained by several K/Ar dates on white mica. As temperatures in the southern Rhenohercynian zone always remained below the white mica closure temperature, the synkinematically grown micas provide formation ages related to deformational peaks (Klügel & Oncken, 1994).

Exhumation was completed at the end of the Carboniferous. Basins like Wetterau trough and Saar-Nahe basin situated between northern Saxothuringian and southern Rhenohercynian zones recorded the sedimentological response to the final stage of erosion and tectonic denudation.

3. Modeling Approach

A two-dimensional thermal-kinematic modeling approach is used to calculate the transient temperature field during crustal thickening by thrusting as well as crustal thinning by erosion and extension. Modeling is based on the finite element code ANSYS (registered trademark of Ansys Inc., Houston). Triangular and rectangular coupled-field elements with thermal-structural field capabilities are used to compute the thermal evolution of a thrust sheet moving over a ramp structure onto the footwall. The fault plane is modeled with contact elements. Friction along contact surfaces and shear heating are not taken into account.

The finite element net represents an area of 600 km width and 32 km depth using a grid point spacing of 6 km in the horizontal and 2 km in the vertical direction. The material parameters used to calculate the crustal temperature

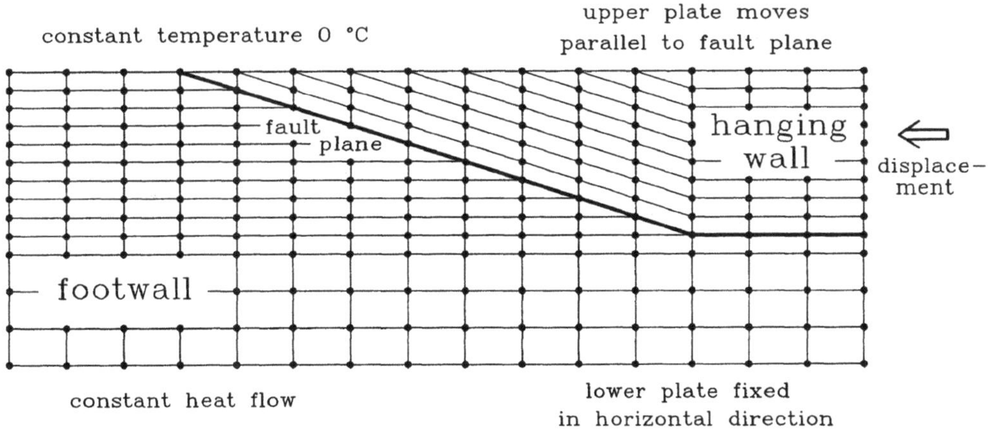

Figure 3. Part of the initial finite element grid.

field and its changes with time are summarized in Table 2. It is assumed that radiogenic heat production is confined to the upper 18 km of crust. The initial temperature distribution is assumed to be in equilibrium with mantle heat flow and upper crustal radiogenic heat production. The contribution by heat flow from the mantle into the crust was increased from NW to SE to account for the geodynamic position of the Spessart complex closer to the magmatic arc, which formed during earlier SE-directed subduction of the Lizard - Giesen - Harz ocean.

K	thermal conductivity	$2.5\ W\ m^{-1}\ K^{-1}$
A	radiogenic heat production	$2.7 \times 10^{-6}\ W\ m^{-3}$
C	heat capacity	$9.0 \times 10^{2} J\ K^{-1} kg^{-1}$
ρ	density	$2.78 \times 10^{3} kg\ m^{-3}$

Table 2. Material parameters used in modeling.

Following Shi & Wang (1987), Germann (1990) and van Wees et al. (1992), the kinematic model is based on a velocity vector in the hanging-wall which is always parallel to the fault plane directly beneath it. The horizontal component of plate convergence is assumed to be constant within the thrust sheet. Consequently, vertical movements result from the dip of the fault while bending at hinge lines is accommodated by vertical shear. Local Airy isostasy is maintained throughout the simulation.

Comparison between modeled P-T-t paths and the petrologic, thermochronological and tectonic calibration data listed in Table 1 are based on a forward modeling approach. For a set of lithospheric parameters (Table 2) the burial and exhumation histories are varied by trial and error until a good fit between observed and modeled data is achieved.

4. Burial, Convergence and Exhumation History

Modeling results are summarized in a number of temperature cross-sections through the suture zone (Fig. 4) as well as P-t and T-t diagrams for each of the three complexes studied (Fig. 5). The initial geodynamic setting of the study area in a fore-arc to foreland position after subduction of oceanic crust was completed is schematically depicted in Fig. 4 a.

Thermal-kinematic modeling of the northernmost Saxothuringian and

Rhenohercynian zones began during the latest Viséan. An initial crustal thickness of 32 km is assumed as the youngest sediments known from the southern Rhenish Massif were deposited in a marine environment. However, it should be noted, that crustal thickness is no parameter of primary importance, because crustal thickening occurred by stacking of upper crustal thrust sheets and radiogenic heat production is assumed to be concentrated in the upper crust. The initial temperature field is represented by the steady-state solution shown in Fig. 4 b. The initial surface heat flow is assumed to increase from 73 to 90 mW m^{-2}. The temperature increase towards the south reflects the geodynamic setting of the study area in a foreland (= previous passive margin of the Rhenohercynian shelf) to fore-arc setting north of a diminishing subduction zone. The temperature - depth relationship in the northern Saxothuringian zone meets average peak metamorphic requirements for the Spessart region, i.e. about 600° C at 21 km. Rocks of the future NPZ and TKU are still at or close to the surface.

Cooling of the northern Saxothuringian zone and deep burial of the southern Rhenohercynian zone occurred almost simultaneously at the Viséan / Namurian boundary. Cooling rates of temporarily up to 70 °C/Ma as documented by the thermochronological data of the Spessart complex can be explained either by rapid exhumation or by rapid thrusting. Exhumation requires average decompression rates of 3.2 mm/a between 326 and 321 Ma, during the Early Namurian. Alternatively, the data presently available can be explained by cooling due to thrusting over colder footwall rocks (see below). Horizontal thrusting velocities of at least 30 mm/a in combination with some minor exhumation would also satisfy the T-t calibration data. It should be noted that the interpretation by rapid exhumation implies significant differential movement between Spessart complex and NPZ, which requires a major thrust zone between the two units. Additional P-T and P-t data will be needed to determine which interpretation is correct.

Deep burial of the southern Rhenohercynian zone is modeled by a single thrust sheet emplacing 18 km of crust over the NPZ. The temperature distribution in the overthrust structure strongly depends on the thrusting velocity and thrust ramp angle (see also Shi & Wang, 1987; Germann, 1990). The lower the ramp angle, the faster the thrust sheet has to move to keep temperatures at a fixed point in the footwall below a certain limit (see Henk, 1995 for details). The upper temperature of 330 °C at 5 - 6 kbar, indicated by the absence of biotite in the NPZ, severly constrains the convergence velocities between the Saxothuringian and Rhenohercynian zones during the Late Viséan and Namurian. For ramp angles of 18° as suggested by DEKORP 2-S, initial thrusting velocities of at least 30 mm/a are required to keep temperatures at 18 km

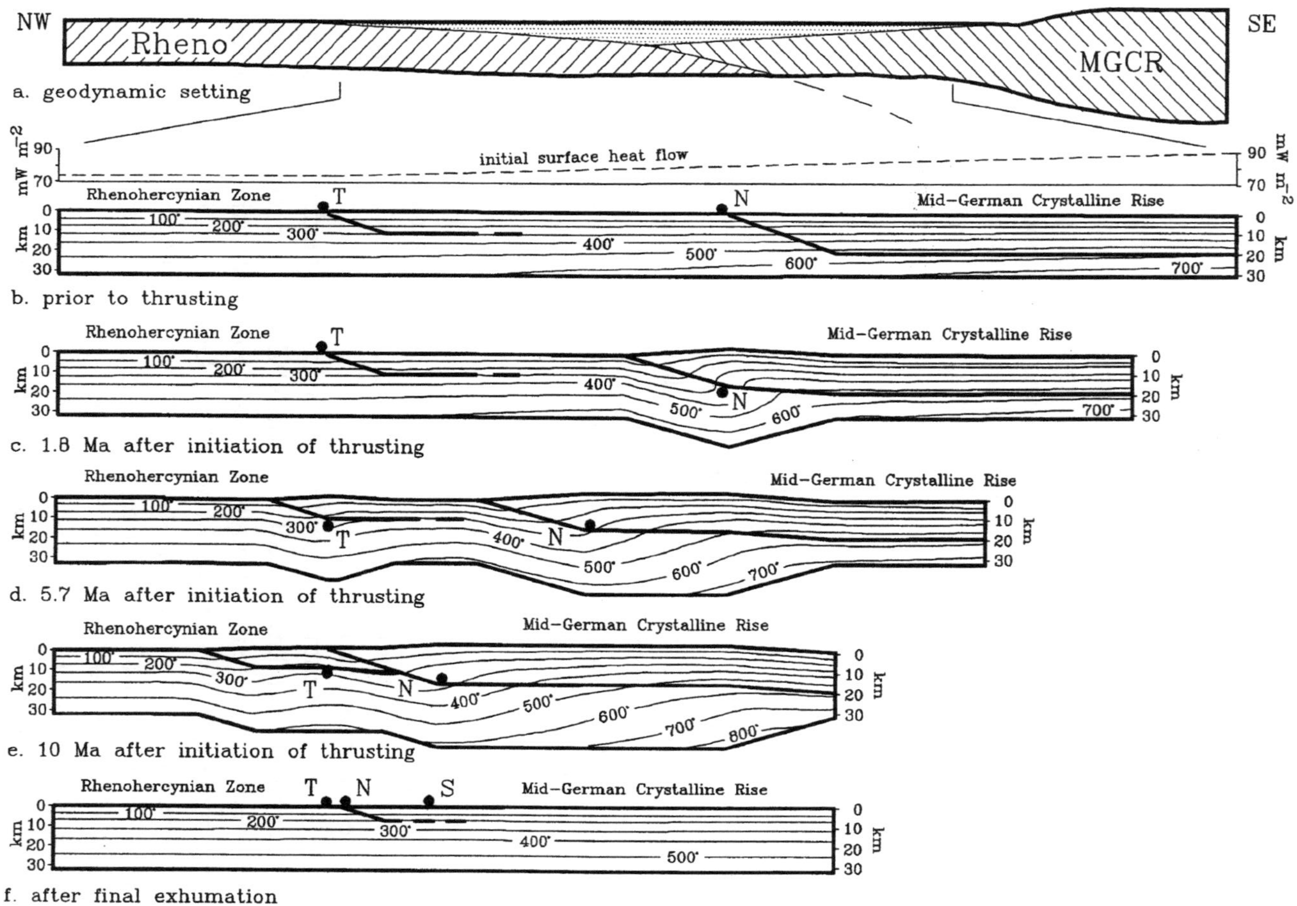

Figure 4. Temperature distribution at the suture between the Rhenohercynian and Saxothuringian zones during Viséan to Stephanian times (see text for further explanation; T = Taunuskamm Unit, N = Northern Phyllite Zone, S = Spessart Complex). No vertical exaggeration.

depth in the footwall below the critical level of 330 °C. (see also Henk, 1995). Any slower convergence would have caused excessive heating, which is not documented in the NPZ.

Maximum burial of the NPZ due to NW-directed thrusting occurred at about 324 Ma. Because of the required thrusting rates, burial to 18 km depth was very rapid indeed and essentially completed within less than two Ma (Fig. 4 c). Consequently, it has to be assumed that thrusting of the northern Saxothuringian onto the Rhenohercynian zone started at about 326 Ma. Rapid convergence and overthrusting strongly reduced the thermal gradient in the middle crust but were not sufficient to cause any significant inversion of the temperature field. As indicated by the pressure data (Klügel & Oncken, 1994), the NPZ remained at approximately 18 km depth until about 310 Ma. If the NPZ would have stayed in its initial position in the thrust's footwall during these 15 Ma, temperatures would have increased to more than 550 °C. In order to prevent thermal equilibration and to keep temperatures below 330°C, the rocks had to be cooled - but without exhumation. An efficient process is required to prevent thermal equilibrium over a time span of several million years. A possible mechanism is cooling from beneath by continuous thrusting over colder footwall rocks. Thus, the following scenario may be envisaged: shortly after the initial burial stage the NPZ was accreted to the base of the hanging wall and subsequently moved with the advancing Saxothuringian zone over the former Rhenohercynian shelf. The narrow temperature interval of 270 - 300 °C, indicated by mineral assemblage and fabric of the shear zone rocks, can be achieved assuming thrusting velocities of 20 mm/a until 310 Ma (Fig. 4 d, e; see also Henk, 1995).

The TKU experienced maximum burial to approximately 11 km depth at about 320 Ma. As suggested by data from the NPZ, convergence between the Rhenohercynian and Saxothuringian zones had already slowed down at that time. In contrast to the NPZ, rocks of the TKU always remained in their footwall position. Thus, the time lag of 10 Ma between thrusting and exhumation was sufficient to achieve almost complete thermal equilibration with the steady-state geotherm (Fig. 4 d, e).

Relative movement between the Rhenohercynian and Saxothuringian zones was not confined to thrusting, but also involved strike-slip displacement, particularly in the final collision stage (Oncken, 1988). It was probably during the transpressional deformation phase at about 310 Ma that the NPZ was partly exhumed and placed adjacent to the TKU. This juxtaposition occurred within the margins of error of the radiometric dates and proceeded almost isothermally. Consequently, for this stage no definite estimate on exhumation velocities can be derived from thermal modeling. Depending on the time assigned, rates

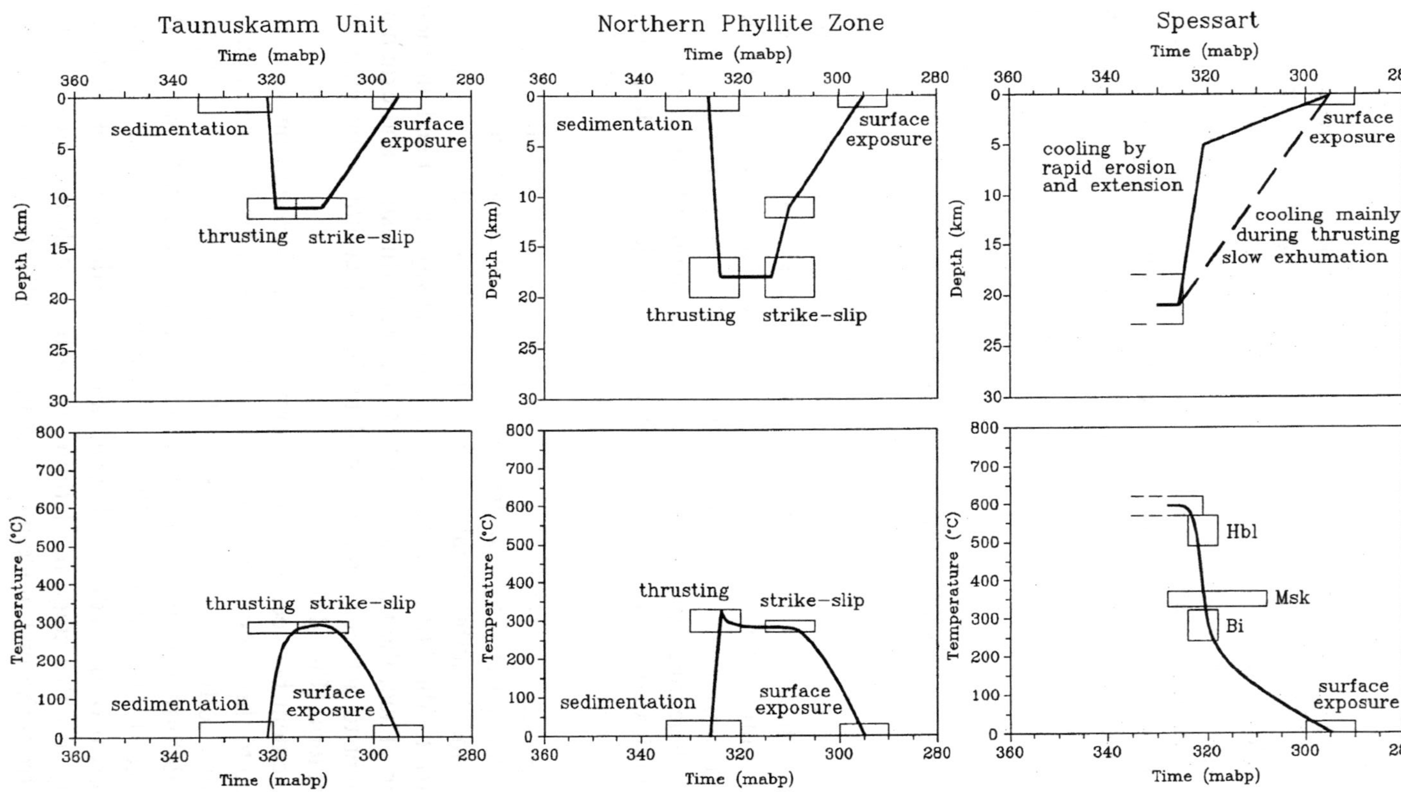

Figure 5. Modeled Temperature - time and Depth - time paths for Spessart Complex, Northern Phyllite Zone and Taunuskamm Unit. Boxes show calibration data including uncertainties in metamorphic conditions, age and mineral closure temperature.

between 1 and 3 mm/a can be expected. The subsequent model path that approximates the common exhumation of TKU and NPZ suggests unroofing at velocities of about 0.70 mm/a between 310 Ma and final surface exposure at about 295 Ma. Exhumation of the study area was completed during the uppermost Carboniferous as already the oldest sediments of the Wetterau trough contain a complete spectrum of detritus derived from Spessart complex, NPZ and TKU (Kowalczyk, 1983).

5. Conclusions

Thermal-kinematic modeling presented in this study provides some quantitative insights into the final convergence history of the Saxothuringian and Rhenohercynian zones and, thus, contributes to a better understanding of the complex Late Variscan evolution in Central Europe. It can be shown, that the thermally unequilibrated rocks of the NPZ record convergence rates during the Late Viséan / Early Namurian head-on collision of the Rhenohercynian and Saxothuringian zones of at least 30 mm/a. The reduced convergence rates during the rest of the Namurian and Westphalian overthrusting indicate that direct contact of the zones decelerated the collision. These convergence rates, derived from thermal modeling, are in broad agreement with independent estimates based on flysch and deformation front migration (Oncken, personal communication 1994).

After accretion to the base of the Saxothuringian tectonometamorphic unit, thermal equilibration of the NPZ was prevented by continuous thrusting over the colder footwall rocks. As the NPZ advanced with the Saxothuringian zone onto the Rhenohercynian zone to the north, its rocks must have been deposited originally much farther SE of their present position (with respect to TKU and Rhenish Massif). It is difficult to estimate this distance because the radiometric data for the southern Rhenohercynian zone have an error of $\pm$ 5 Ma. The modeled convergence rate for this stage is equivalent to 20 km displacement for every million years of thrust movement. Thus, a conservative estimate of 10 million years duration between thrusting and juxtaposition next to the TKU would imply about 200 km northwestward movement of the NPZ relative to its initial position.

Exhumation during transpressional movements in the southern Rhenohercynian zone and possibly also during the rapid cooling stage of the northern Saxothuringian zone occurred at rates of up to 3 mm/a. Decompression at such velocities is difficult to explain by erosion only and rather suggests a substantial contribution of tectonic denudation. However, exhumation of the study area was partly contemporaneous with continuous compression in the Saxothuringian basin to the south (Franke & Oncken, 1990). Thrusting in the Rhenohercynian zone to the north continued until Westphalian D (Ziegler, 1990). Extensional strain orthogonal to the collision direction, i.e. lateral extrusion, could be an adequate process to explain the regional Late Variscan convergence as well as the contribution of tectonic denudation to the modeled exhumation rates. Preferential sites for lateral extrusion and, consequently, efficient decompression are thrust ramps. Seismic profile DEKORP 2-S suggests, that comparable fault geometries occur abundantly in the upper crust of the study area (Fig. 2; Behr & Heinrichs, 1987).

The study shows, that during Late Variscan times significant crustal thickening and exhumation occurred almost contemporaneously and in close spatial relationship at the suture between the Rhenohercynian and Saxothuringian zones. This is in agreement with observations from currently active orogens, which also show, that both processes can act concomittantly in one orogen and that deep burial can pass into uplift in 5 million years or less (Jamieson & Beaumont, 1989). Recognition of such fast successions of various geodynamic processes is at or beyond the resolution of most traditional P-T-t or stratigraphic data. More refined thermobarometric and geochronological data are required to gain a more detailed understanding of orogens and their evolution through time.

6. Acknowledgements

The author gratefully acknowledges discussions with O. Oncken, M. Okrusch and T. Klügel. Comments by an anonymous reviewer helped to improve the manuscript. This research was generously funded by a grant from the Deutsche Forschungsgemeinschaft as part of the priority program "Orogenic processes - their simulation and quantification using the Variscides as example".

7. References

Anderle, H.-J., Massonne, H.-J., Meisl, S., Oncken, O. & Weber, K. (1990): Southern Taunus Mountains. - Field Guide "Mid-German Crystalline Rise & Rheinisches Schiefergebirge", Int. Conf. on Paleozoic orogens in Central Europe, Göttingen-Giessen, Aug. - Sept. 1990: 125-148.

Behr, H.J. & Heinrichs, T. (1987): Geological interpretation of DEKORP 2-S. A deep reflection profile across the Saxothuringian and possible implications for the Late-Variscan structural evolution of Central Europe. - Tectonophysics, 142: 173-202.

Franke, W. & Oncken, O. (1990): Geodynamic evolution of the North-Central Variscides - a comic strip. - *In*: Freeman, R., Giese, P. & Müller, S. (eds.): The European Geotraverse: integrative studies. Results from the Fifth Study Center. - European Science Foundation: 187-194.

Franke, W. (1992): Phanerozoic structures and events in Central Europe. - *In*: Blundell, D., Freeman, R. and Mueller, S. (eds.): A continent revealed - the European geotraverse. - Cambridge University Press: 164-180.

Germann, D. (1990): Thermische und strukturelle Prozesse in der Überschiebungstektonik und ihre numerische Behandlung. - PhD thesis, Techn. Univ. Clausthal: 219 p.

Henk, A. (1995): Late Variscan exhumation histories of the southern Rhenohercynian Zone and western Mid-German Crystalline Rise - results from thermal modeling. - Geol. Rdsch., 84: 578-590.

Jamieson, R.A. & Beaumont, C. (1989): Deformation and metamorphism in convergent orogens: a model for uplift and exhumation of metamorphic terrains. - *In*: Daly,J.S., Cliff, R.A. & Yardley, B.W.D. (eds.): Evolution of Metamorphic Belts. - Geol. Soc. Spec. Publ. No. 43: 117-129.

Klügel, T. & Oncken, O. (1994): Struktur und Kinematik einer variszischen Suturzone am Beispiel des Taunus-Südrandes. - Göttinger Arb. Geol. Paläon., Sb1: 179-181.

Kowalczyk, G. (1983): Das Rotliegende zwischen Taunus und Spessart. - Geol. Abh. Hessen, 84: 99 p.

Nasir, S., Okrusch, M., Kreuzer, H., Lenz, H. & Höhndorf, A. (1991): Geochronology of the Spessart Crystalline Complex, Mid-German Crystalline Rise. - Mineralogy and Petrology, 44: 39-55.

Okrusch, M. (1990): Metamorphism in the Odenwald and Spessart crystalline mountains (Mid-German Crystalline Rise). - Field Guide "Mid-German Crystalline Rise & Rheinisches Schiefergebirge", Int. Conf. on Paleozoic orogens in Central Europe, Göttingen-Giessen, Aug. - Sept. 1990: 81-91.

Oncken, O. (1988): Geometrie und Kinematik der Taunuskammüberschiebung - Beitrag zur Diskussion des Deckenproblems im südlichen Schiefergebirge. - Geol. Rdsch., 77: 551-575.

Shi, Y. & Wang, C.-Y. (1987): Two-dimensional modeling of the P-T-t paths of regional metamorphism in simple overthrust terrains. - Geology, 15: 1048-1051.

van Wees, J.D., Jong, K. de & Cloetingh, S. (1992): Two-dimensional P-T-t modelling and the dynamics of extension and inversion in the Betic Zone (SE Spain). - Tectonophysics, 203: 305-324.

Ziegler, P.A. (1990): Geological Atlas of Western And Central Europe. - Shell Internationale Petroleum Maatschappij, 2nd ed.: 239 p.

MYLONITIC FOLIATION DEVELOPED BY HETEROGENEOUS PURE SHEAR UNDER HIGH-GRADE CONDITIONS IN QUARTZOFELDSPATHIC ROCKS (CHÍMPARRA GNEISS FORMATION, CABO ORTEGAL COMPLEX, NW SPAIN)

FRANCISCO J. FERNANDEZ and ALBERTO MARCOS
Departamento de Geología. Universidad de Oviedo.
c/ A. de Velasco s/n, 33005 Oviedo, Spain

Abstract

The quartzofeldspathic Chímparra Gneiss Formation contains S-type mylonites developed under high pressure-high temperature metamorphic conditions. These mylonites may appear either as an homogeneous planar foliation that obliterates the previous textures and structures of the rock, or as an inhomogeneous planar foliation forming anastomosing shear bands. In the anastomosing areas, quasi-lenticular pods of migmatitic gneiss have been partially preserved from later deformation and show a previous migmatitic texture. Within these pods, the migmatitic banding has been symmetrically folded and the geometry of the folds indicates a progressive flattening towards the edges of the pods. The mylonitic foliation is parallel to the axial planes of these folds, and causes an intersection lineation parallel to the fold axis; however, a mineral lineation is absent. Kinematic criteria deduced from the geometry of polycrystalline quartz aggregates, extensional crenulation cleavages (ECCs), and quartz c-axis fabrics do not indicate any consistent preferred sense of movement. Both macro- and microstructural analyses suggest an inhomogeneous pure shear or flattening as the main deformation process related to the development of the mylonitic foliation.

1. Introduction

The Cabo Ortegal Complex, together with other isolated outcrops of *HP-HT* rocks preserved in the NW of the Iberian Peninsula, constitute the remnants of a lower crustal and mantle slice emplaced onto the continental margin of Gondwana during the Variscan orogeny (Ries and Shackleton, 1971; Marcos et al., 1984; Pérez-Estaún et al., 1991). Metaultrabasic, metabasic and quartzofeldspathic rocks in the Cabo Ortegal Complex (Fig. 1) preserve an early *HP-HT* metamorphism, and record a subsequent retrogression to greenschist facies conditions (Vogel,1967; Engels, 1972; Gil Ibarguchi and Arenas, 1990).

O. Oncken and C. Janssen (eds.), Basement Tectonics 11, 17–34.

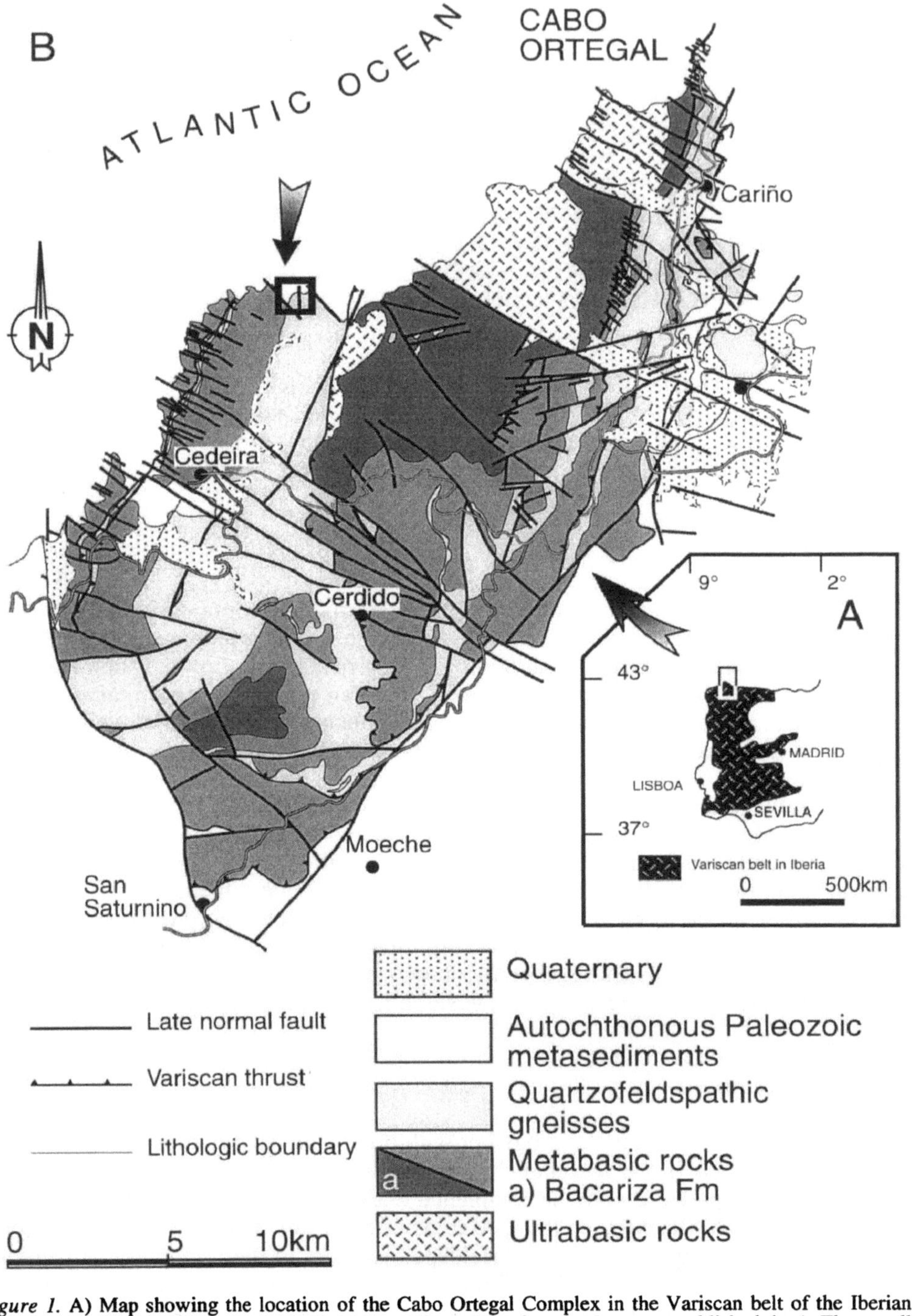

Figure 1. A) Map showing the location of the Cabo Ortegal Complex in the Variscan belt of the Iberian Peninsula. B) Geological map of the Cabo Ortegal Complex (after A. Marcos, unpublished data). The small black box in "B" indicates the study area.

One of the most outstanding structural features found in the Cabo Ortegal rocks is a generalized S-type mylonitic foliation that started to develop under HP granulite facies conditions and which subsequently re-equilibrated under amphibolite facies conditions (Fernández, 1993). This foliation was subsequently folded and thrusted during emplacement of the Cabo Ortegal Complex onto low metamorphic grade Paleozoic rocks (Marcos et al., 1984; Arenas et al., 1986). During this last exhumation process almost all of the minerals that defined the foliation were retrogressed to greenschist facies conditions.

The purpose of this paper is to analyze the S-type mylonitic foliation developed in the Chímparra Gneiss Formation (Vogel, 1967) during the early stages of its tectonic history under lower crustal metamorphic conditions. S-type mylonitic foliations are not very frequently described in the literature relating to the deformation of lower crustal rocks. The Tarroiba cliff area (Fig. 1) was chosen for this study because the mylonitic foliation has largely spared from folding and thrusting. In this investigation the processes that produced this mylonitic foliation were studied in order to better understand the tectonic exhumation and emplacement of lower crustal slices into upper crustal levels.

The Chímparra Gneiss Fm consists mainly of garnet and kyanite quartzo-feldspathic gneisses. It also contains inclusions of eclogite and mafic granulites (Vogel, op cit.; Gil Ibarguchi et al., 1990), although in the Tarroiba section only some boudins of calc-silicate rocks can be found. *PT* conditions estimated from these gneisses by garnet-biotite thermometry and garnet-plagioclase thermobarometry give maximum values of 818 ± 65 °C and 1579 ± 176 MPa and minimum values of 597 ± 65 °C and 903 ± 220 MPa, respectively (Fernández, in press). The maximum values correspond to the earliest stages of development of the mylonitic foliation and the minimum values to the re-equilibration under amphibolite conditions. Geochemical data point towards a metasedimentary origin for the Chímparra quartzofeldspathic gneiss (Vogel, 1967; Fernández, in press).

2. Mylonitic foliation and macrostructures

The Tarroiba Cliff is a 400 m long and 200 m high outcrop of the heterogeneously mylonitized Chímparra gneiss. The mylonitic foliation appears in the outcrop 1) is a homogeneous planar foliation, that obliterates the previous structures and textures of the rock, in a 100 m wide southwestern zone near the contact with the neighbouring metabasic rocks (Fig. 1) and 2) is an inhomogeneous planar foliation forming anastomosing shear bands that isolate meter- scale lens-shaped pods of weakly deformed gneisses (Fig. 2), in a ca. 300 m wide northeastern area. The interiors of the pod exhibits a migmatitic structure characterized by alternating non-planar, leucocratic and melanocratic bands.

Figure 2. Field photograph of a lens-shaped pod bounded by ductile shears. Bar length is 1 m. View is looking towards WSW.

A similar texture, characterized by quartzofeldspathic and phyllosilicate-rich domains, can be seen in thin sections of the mylonites from both areas. The quartzofeldspathic domains contain prismatic crystals of quartz, plagioclase, garnet and kyanite which define the planar foliation. Only quartz and plagioclase show evidence of dynamic recrystallization (i.e. the development of ribbons, new grains formed by migration and rotation) and annealing. The phyllosilicate-rich domains contain biotite, white mica and garnet which define an anastomosing texture characterized by the development of conjugate ECCs. The occurrence of prismatic garnets within the white mica phenoblasts, oriented parallel to the mylonitic foliation, are the main evidence for *HP* metamorphism coinciding with the development of the mylonitic foliation.

The area displaying a planar homogeneous foliation exhibits a widespread retrogressive metamorphism post-dating the development of the mylonitic foliation. It is characterized by the partial replacement of garnet by biotite, phengite by muscovite, plagioclase by sericite and biotite by chlorite.

This study focussed on the analysis of the area containing the anastomosing foliation where the earliest stages of its evolution are preserved. In this area, the mineral composition of weakly sheared rock in the interior of the pods is similar to that in the mylonitized zones or shear bands. Furthermore, there are no significant differences in major, trace and REE elements between the central parts of the pods and the surrounding mylonites (Fernández, in press), suggesting that the formation of the mylonitic foliation took place under fluid-free conditions. The interiors of the pods exhibit a migmatitic texture characterized by alternating non-planar leucocratic and melanocratic bands.

2.1. Lenticular pods

The pods are three-dimensional bodies of meter-scale whose geometry can be defined by the axial ratio Ra/b and Rb/c in which a, b and c are the major, intermediate and minor axes of the body, respectively. Only in some exceptional cases can the three axes of a pod be directly measured in the cliff (Fig. 3).

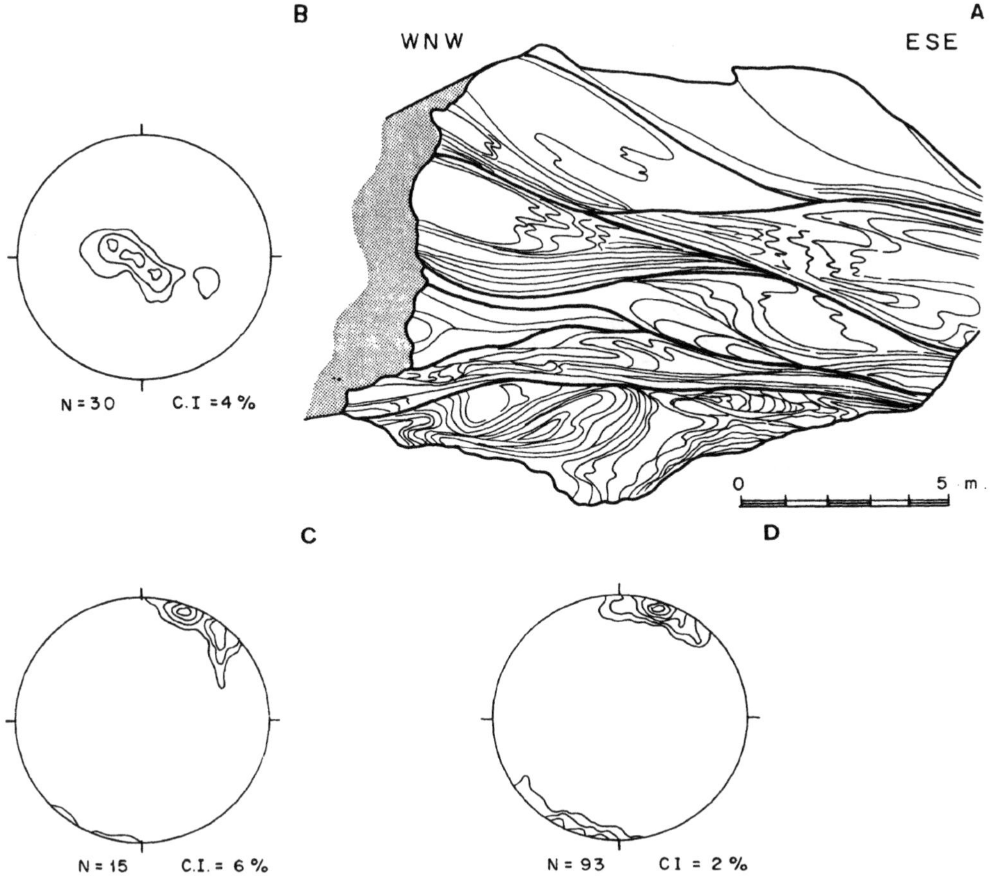

Figure 3. A) Field sketch of the anastomosing foliation and pods of weakly sheared migmatitic gneisses. Stereographic plots (equal area) of B) mylonitic foliation. C) intersection lineation between the foliation and the composition banding and D) axes of minor folds (see explanation in the text).

The axial ratio has been determined for six pods and the values plotted in diagram of Fig. 4. Ra/b values vary from 1 to 1.5 and Rb/c from 2.5 to 4. In order to obtain a representative geometry for all pods, a geometrical method has been developed which uses a polar projection of foliation planes measured throughout the outcrop. From this projection the major and middle axes of an ideal pod, assuming a perfect lens shape, can be determined and the minor axis can be obtained from a Mohr circle construction. The average ideal pod has axial ratios for Ra/b of 1.46 and for Rb/c of 2.14 (Fig. 4). From the axial ratios it can be deduced that the geometry of the pods correspond to bodies with a shorter axis (c)lying in the direction normal to the foliation plane. The pods are slightly elongated along the direction N15E.

It is important to note that the axial ratios are geometric relationships which do not correspond to measures of strain. The lenticular bodies do not result from deformation of existing bodies of rock; they form during deformation due to the anastomosing pattern of the shear bands. The quantification of the three-dimensional shape of the pods allows the geometry of weakly deformed rock to be defined, and indicates which processes generated the foliation.

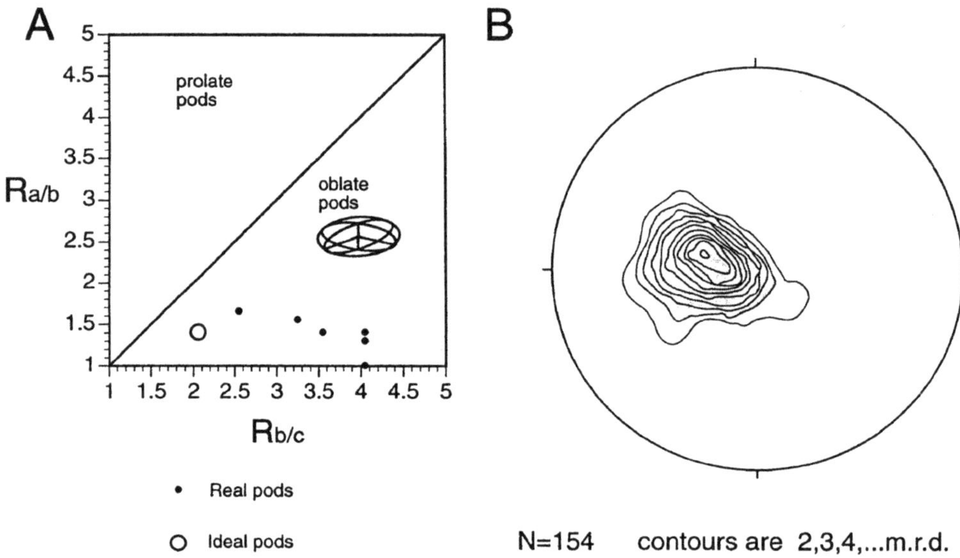

Figure 4. A) Diagram showing Ra/b and Rb/c axial relationships of the pods. B) Equal area stereographic proyection of foliation poles (contoured area defines the pod geometry).

2.2. Folds and lineations

The migmatitic foliation preserved within the pods is deformed by centimeter to meter-scale folds. The mylonitic foliation is axial planar to these folds. Fold geometries vary from the pod cores to their edges and into the shear bands. Three different fold geometries can be distinguished between the pods and surrounding shear bands. Near the center of the pods, the folds (Folds I) are open with straight hinges and interlimb angles of 90 - 150°. They are not fully cylindrical due to the irregular thickness of the migmatitic layers. The amplitude/half wavelength relationship is 0.13. Microfolds are also present with similar characteristics. Towards the edges of the pods, the folds (Folds II) are tighter with a larger amplitude and are almost overturned. The amplitude/half wavelength relationship varies from 7.5 - 9 and the hinges are slightly curved. Within the shear bands intrafolial folds with disrupted limbs (Folds III) are found (Fig. 5). The three types of folds have periodicity. In most cases, the folds possess symmetric limbs although S and Z asymmetries are also found.

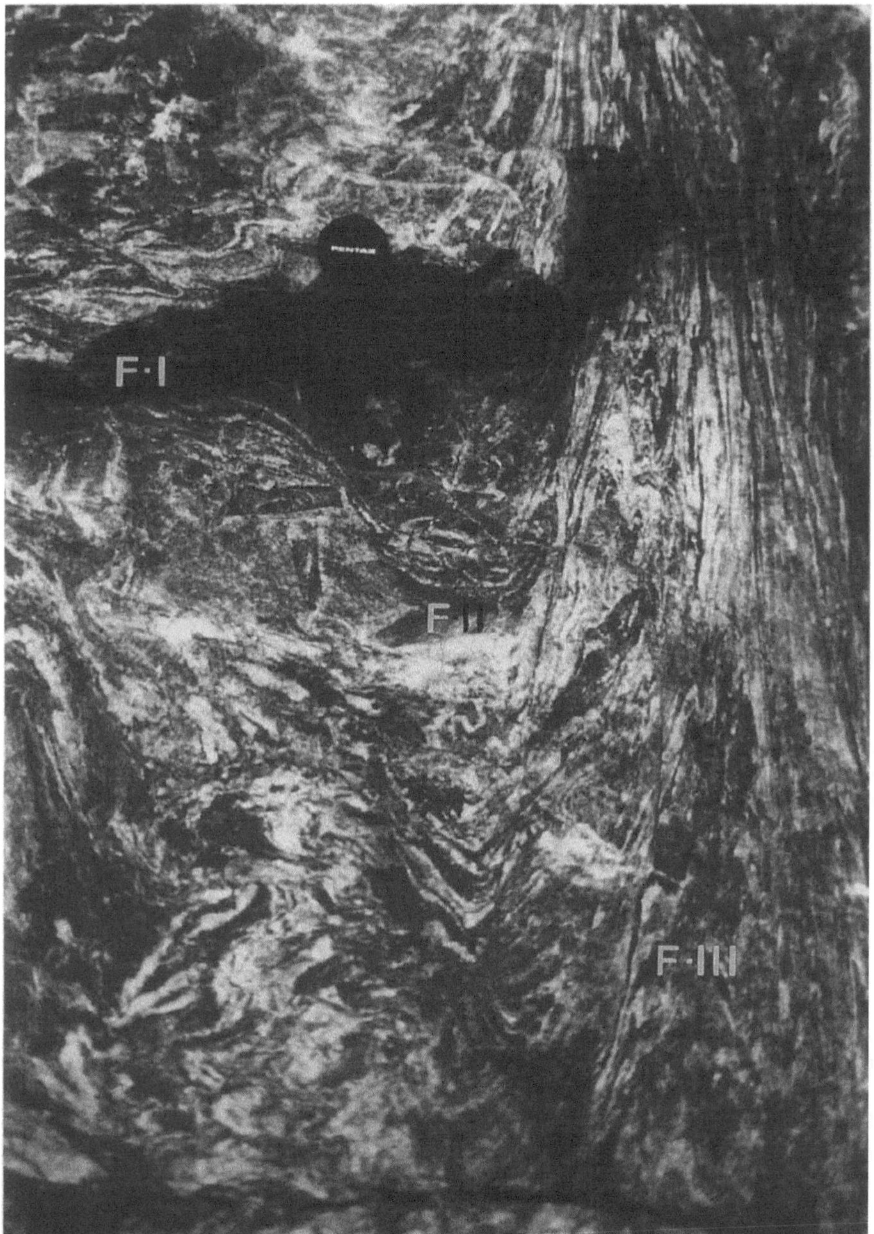

Figure 5. Field photograph of a pod showing variations in the fold geometry from open (F-I) to tight or isoclinal (F-II and F-III) in the neighbouring shear band. View looking up-plunge. Camera lens cap is 50 mm in diameter.

Initially disharmonic migmatitic folds become more harmonic and symmetric in highly strained bands. According to Hudleston's (1973) classification, the folds have a variable amplitude and a sinusoidal D-type shape (Fig. 6). The fold classification based on layer thickness variations (Ramsay, 1967) shows that the fold shape varies from Class 1B to Class 3 from the centers of the pods to the edges. The analysis of the fold population according to the method of Bastida (1993) shows a more scattered plot for Folds I than for Folds II and III (Fig. 7), suggesting a deformation gradient increasing from Folds I in the core of the pods to Folds III in the shear bands.

These data indicate that the folds have been formed by a buckling mechanism with a superposed flattening during the formation of the mylonitic foliation. The difference in fold style (i.e. from buckle folds in the centers to similar folds in the edges) indicates an increasing flattening strain from the centers of the pod to their edges. Although the folds become more harmonic and symmetric with increasing strain is not possible to infer from these data that the folds developed by flexural slip in the low-strain areas and a simple shear mechanism toward the pod rims. In addition, evidence of rotation in the fold limbs, which would be expected for the flexural slip mechanism, is found only very rarely.

A widespread N15E-trending intersection lineation, between the foliation and the composition banding, oriented parallel to the fold axes (Fig. 3) can be found in the outcrops. No mineral orientation has been observed either in the field or in thin section.

3. Mylonitic foliation: microstructure

Different microstructures such as quartz ribbons, extensional crenulation cleavages and quartz c-axis fabrics, together with textures in garnets and other microstructures (such as mechanical twins and deformation lamellae in plagioclase, kinks in phyllosilicates and mica fish) developed during the formation of the mylonitic foliation. They record a major structural change from migmatitic gneiss to mylonitic gneiss.

3.1. Quartz ribbons

Type 2b and 3 quartz ribbons (Boullier and Bouchez, 1978) of varying size occur. They are present on XZ an YZ sections suggesting that the ribbons are flat features. The size of the ribbons relates to both lithology and deformation. In the homogeneous quartzofeldspathic gneiss the quartz ribbons are larger than in the heterogeneous gneiss with phyllosilicate layers (Fig. 8a), and the areas of weaker deformation contain only small quartzofeldspathic lenticular aggregates (Fig. 8b) and no ribbons *sensu stricto*.

The lenticular aggregates usually have symmetrical tails on sections normal to the mylonitic foliation. In some cases asymmetrical tails are also found. Analysis of the orientations of tails reveals a scattered distribution. The ribbons exhibit dynamic recrystallization textures, with the growth of new grains over subjoints normal to the mylonitic foliation, and also over grain edges (Fig. 8A, sample T-21). The distribution of orientations of new grains is usually weakly orthorhombic with a c-axis maximum close to Y. This suggests that prismatic slip was activated in the small grains. The values of differential stress, obtained from the recrystallized grain size of quartz vary from 10 to 70 MPa. These values are lower than expected for rocks of this type (White, 1979) and suggests widespread overgrowing.

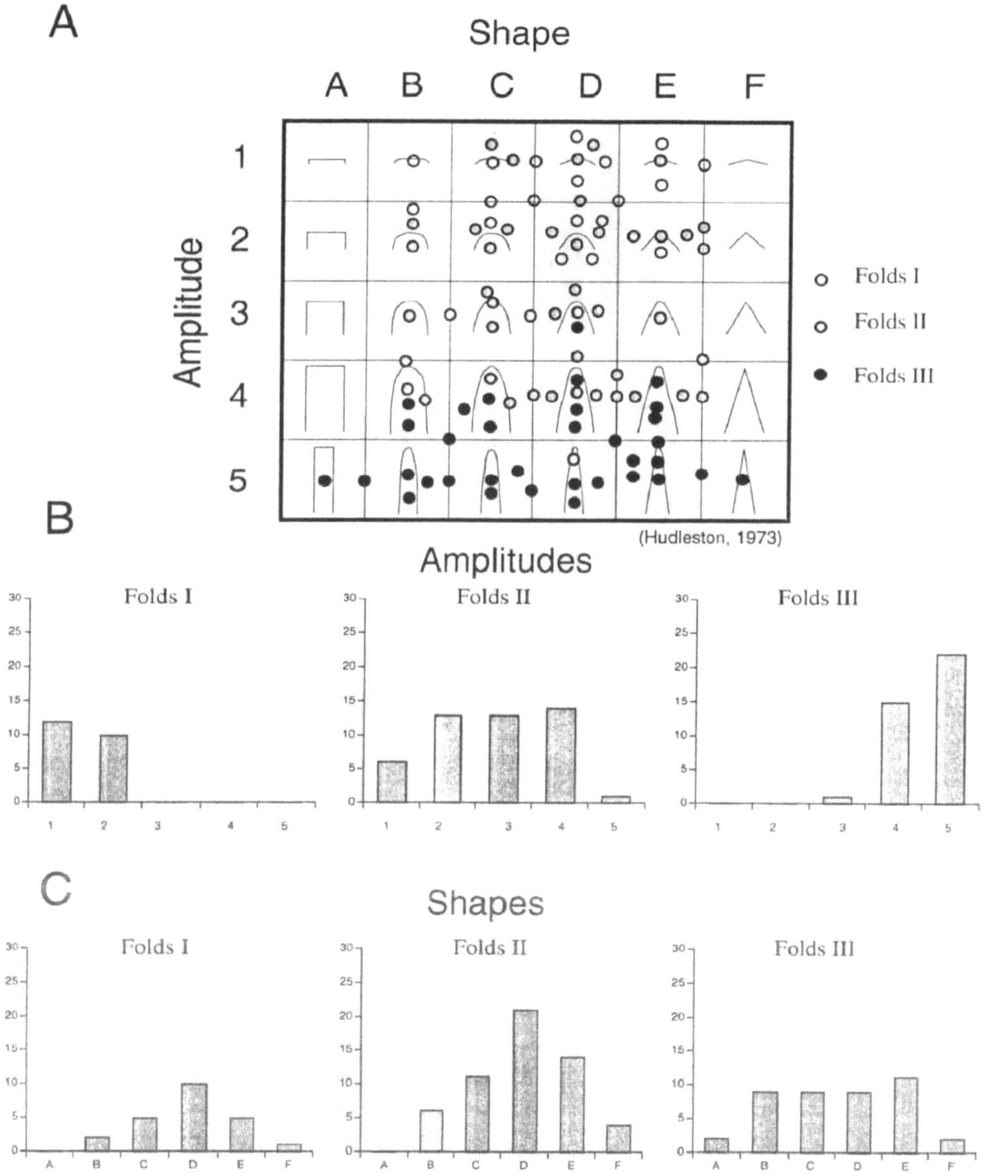

Figure 6. A) Classification of folded surfaces by the Hudleston (1973) method. Histograms of fold amplitudes B) and shapes C) (see text for explanation).

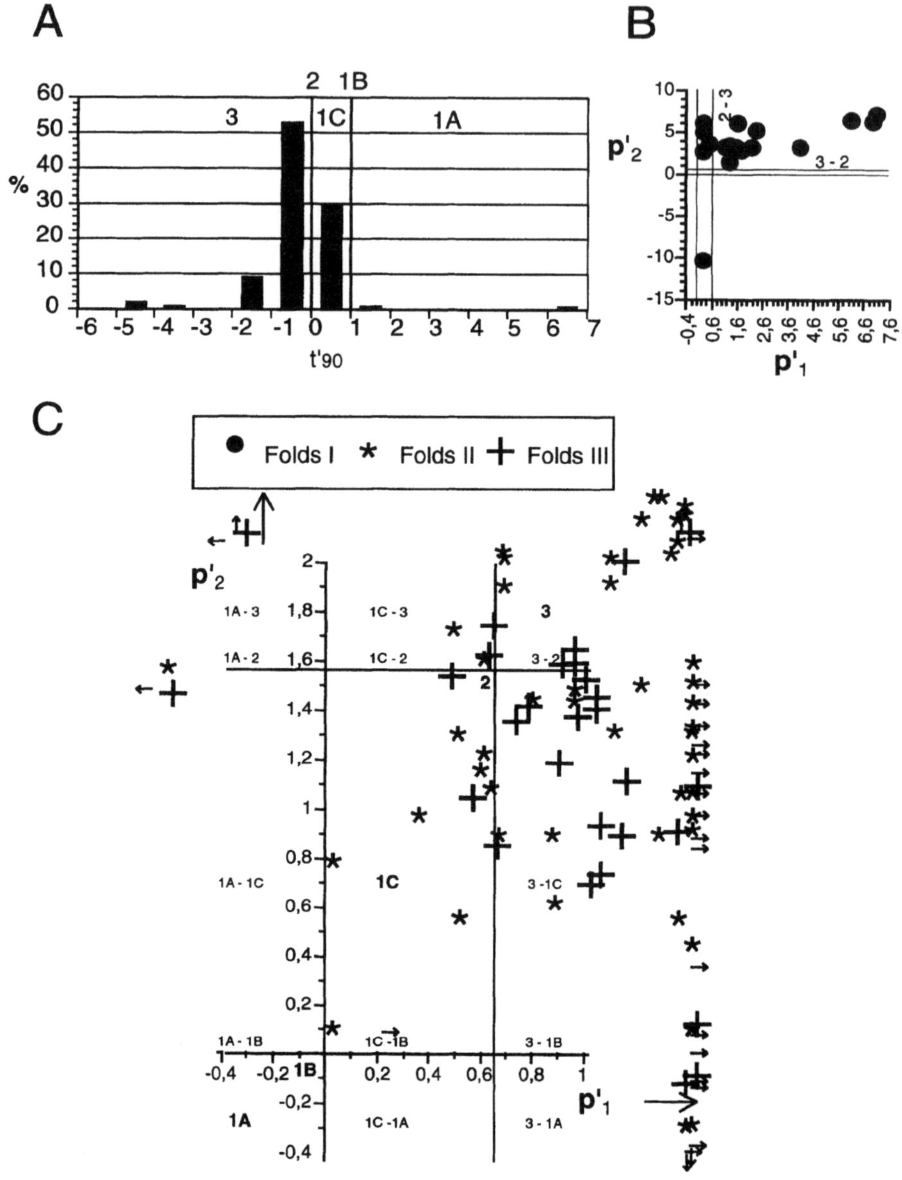

Figure 7. Fold classification according to Bastida (1993). A) Histogram of intercept values (t'_{90}) showing that most of the folds are of Class 3, and some near Class 2. B) Most of the Folds I lie within Class 3 and Class 1C fields in the p'_1-p'_2 diagram. C) Folds II and III plot within Class 3 and Class 3-1C fields in the p'_1-p'_2 diagram.

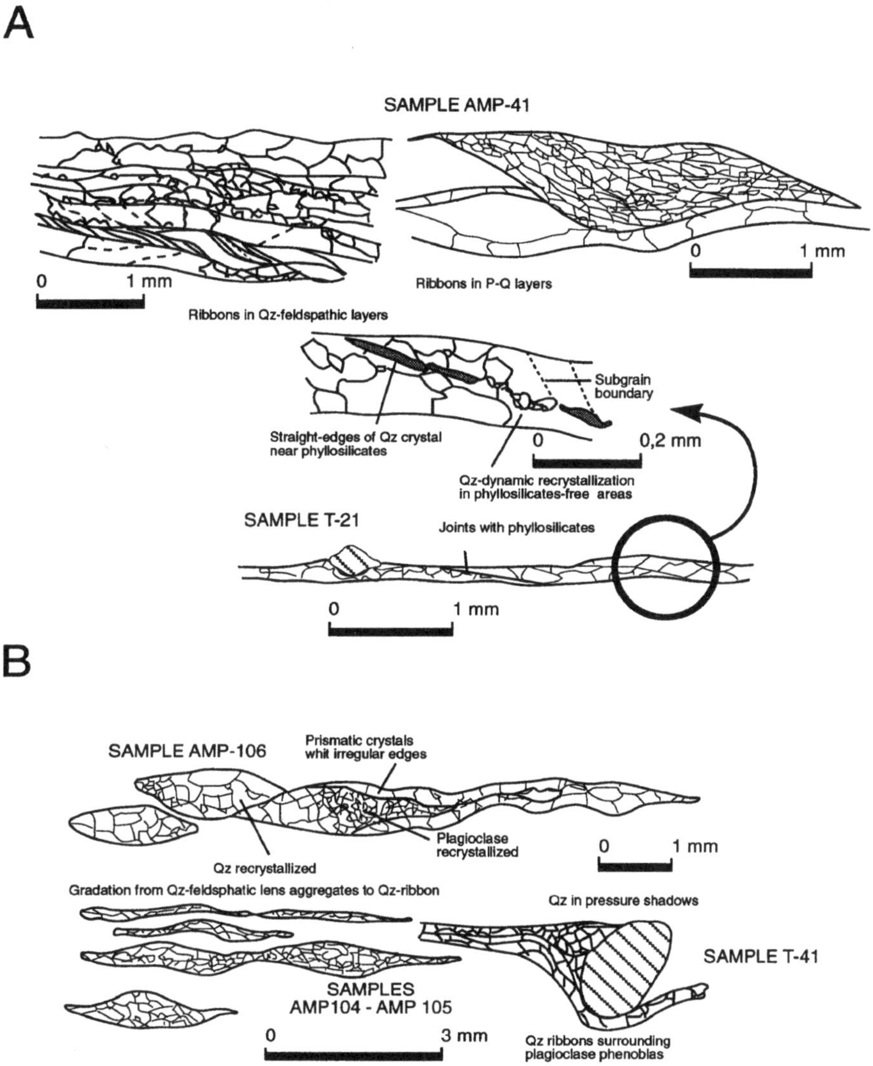

Figure 8. Microscope sketches of A) quartz ribbons in homogeneous quartzofeldspathic gneisses, and B) quartzofeldspathic lenticular aggregates.

3.2. Cracks and flattened shapes of garnets

Flattened garnets are common within the shear bands in the Chímparra Gneiss. According to Ji and Martignole (1994) flattened garnets form by plastic deformation at temperatures above 800 °C and record a component of coaxial strain. Similarly, high temperature conditions are inferred for the mylonitic foliation in the Chímparra gneisses. In thin section, the garnets of the Chímparra mylonites contain cracks oriented at high angles to the mylonitic foliation (Fig. 9A,D). The flattening and cracks in the garnets were controlled by the orientation of σ_1. Using K-compositional mapping on an SEM, two types of cracks were differentiated, types I and II (nomenclature of Prior, 1993). Type I cracks are small and traverse only part of the garnet crystals. These cracks are interpreted as syntectonic because they cut through the compositional zones and near the cracks the zonation has undergone more extensive retrogressive metamorphism than the non-cracked areas of the crystals (Fig. 9B,D). The larger Type II cracks completely cut across the crystals and are filled by recrystallized K-rich phyllosilicate (Fig. 9C) and are, therefore, interpreted as post-tectonic.

3.3. Extensional crenulation cleavages

Extensional crenulation cleavages (ECCs) are found in the areas of high strain within the mylonitic gneiss are better developed in phyllosilicate domains than in quartzofeldspathic domains. They appear as conjugate sets in thin section. The analysis of shear direction from thin sections of oriented samples indicate a predominant direction of shear towards the north but south, east and west orientations are also found (Fig.10), suggesting that the cleavages continued to develop between deformation episodes. Analysis of kinematic indicators, such as σ and δ asymmetries (Driessche and Brun, 1987) in quartzofeldspathic aggregates at macro and microscopic scale, does not reveal any preferred shear direction. This indicates that the early asymmetric shapes were removed by a superposed deformation, as often occurs in areas that have undergone polyphase deformation. However no evidences for reverse movements has been found and the more simple interpretation for the lack of asymmetric indicators is that the deformation ellipsoid has an $\lambda_1/\lambda_2 = 1$ shape.

The presence of δ-type clasts would argue strongly for inhomogeneous, non-coaxial simple shear with a component of spin (Hanmer and Passchier,1991). Nevertheless, in the absence of a preferred shear direction, only an inhomogeneous deformation can be postulated.

3.4. Quartz c-axis fabrics

In the study section, quartz c-axis fabrics seem to be independent of macrostructure. In fact, samples taken from areas preserved from mylonitization (the inner parts of the pods), anastomosing shear bands and homogeneously distributed foliation, record similar c-axis fabrics (Fig. 11). Most samples contain c-axis fabrics with maxima at peripheral positions indicating a basal slip mechanism. Samples T-11.1 and T14.1 (see Fig. 11) record maxima also in a central position, indicating prism <a> glide. The c-axis fabrics have an orthorhombic symmetry with a small-circle girdle in which the half-opening angles vary between 20° and 40°. These fabrics correspond to type-I crossed girdle fabrics (cf. Tullis et al. 1973; Lister and Hobbs 1980) caused by a coaxial deformation.

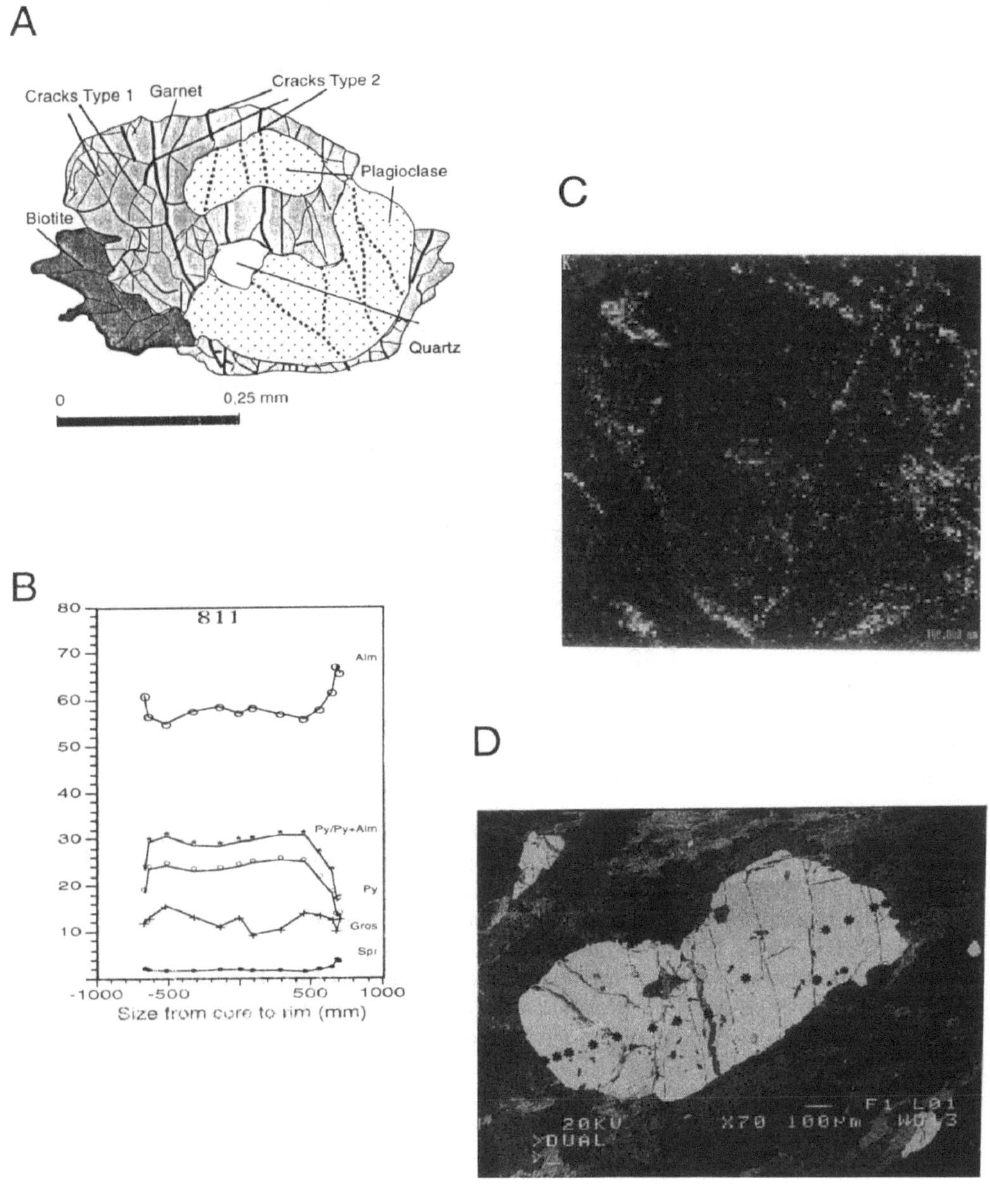

Figure 9. A) Microscope sketch of a fractured garnet . B) Microprobe analyses from a section of a garnet. C) SEM image (resolution of 256 pitxels and 3 scans of 10 min.) of a semi-quantitative K-compositional mapping of the garnet represented in A. D) back-scatter image showing points of microprobe analysis of sample 811 .

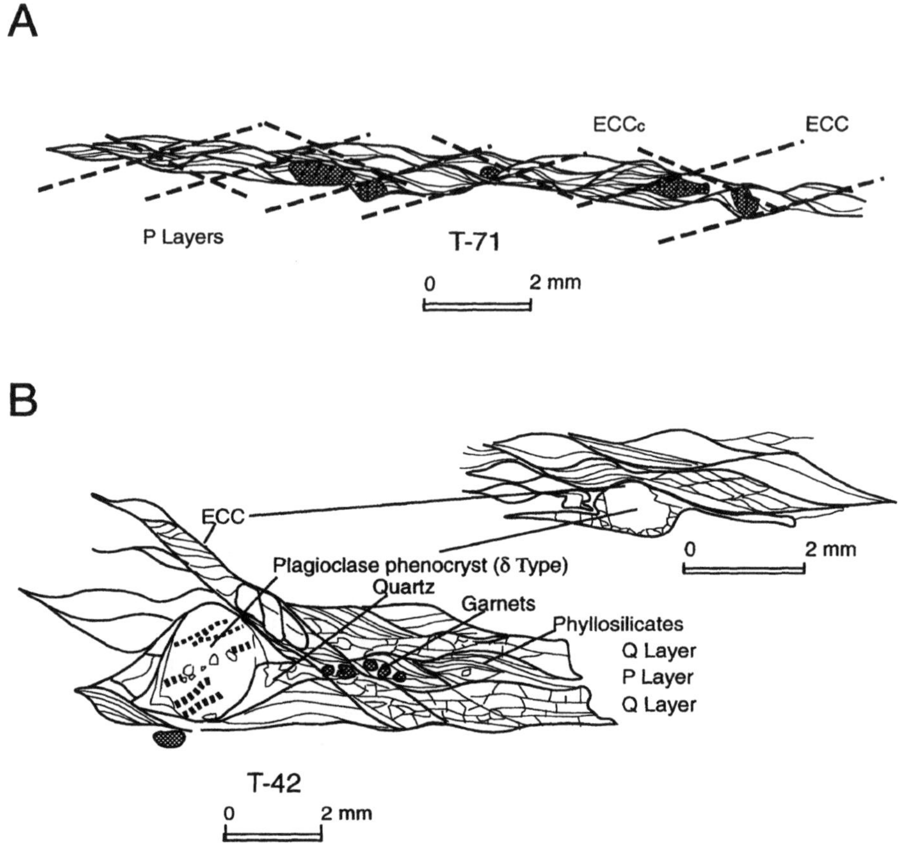

Figure 10. Microscope sketch of A) a symmetrical ECC shear band in a phyllosilicate domain, and B) irregular ECC in quartzofeldspathic domains.

C-axis girdles similar to those described from Tarroiba cliff have been interpreted as characteristic of low-temperature deformation (Nicolas and Poirier 1976; Bouchez and Pecher 1981). However, in the *PT* conditions mentioned above, the anomalous mechanical behaviour of quartz may be due to the absence of fluids such as water (Blacic, 1975; Law, 1990). Samples T-11.1, T-14.1 and T-12.1 (see Fig. 11), are nearest to the contact with the metabasic rocks (see also Fig. 1),showing evidence of greater retrogressive metamorphism, and appear to have deformed by prismatic slip mechanisms. They also show a slight asymmetry indicating simple shear deformation.

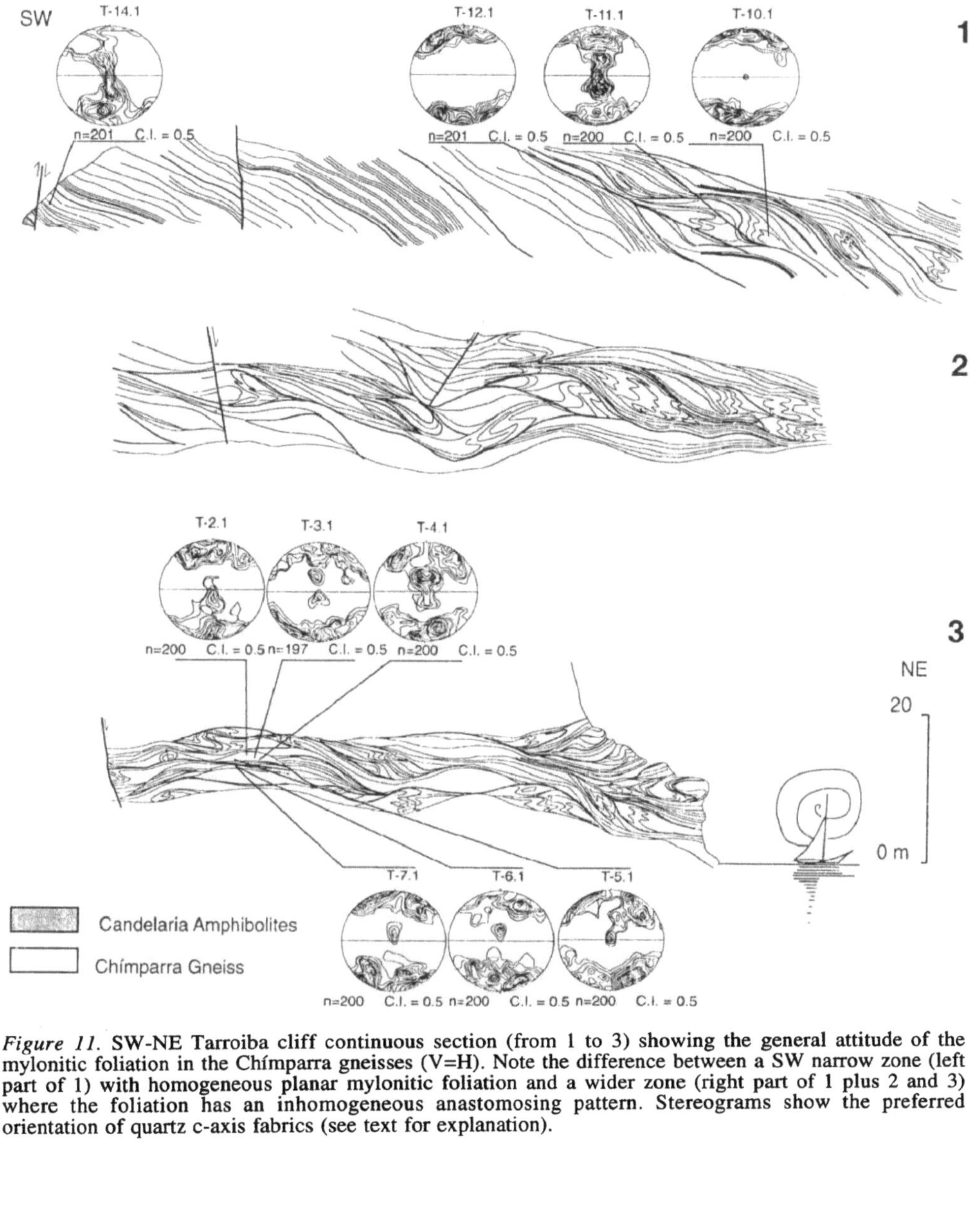

Figure 11. SW-NE Tarroiba cliff continuous section (from 1 to 3) showing the general attitude of the mylonitic foliation in the Chímparra gneisses (V=H). Note the difference between a SW narrow zone (left part of 1) with homogeneous planar mylonitic foliation and a wider zone (right part of 1 plus 2 and 3) where the foliation has an inhomogeneous anastomosing pattern. Stereograms show the preferred orientation of quartz c-axis fabrics (see text for explanation).

4. Discussion and conclusions

The Chímparra Gneiss contains S-type mylonites (Fig. 12) which consist of areas with anastomosing shear bands and areas of homogenous planar foliation. In the anastomosing areas, pods of migmatitic gneiss have been partially preserved from the deformation and show a quasi-lenticular shape with the short axes parallel to the Z direction. The geometries of the folds within the pods show a progressive change from open buckle folds near the centers of the pods to progressively flattened symmetric folds towards their edges. The mylonites possess a strong intersection lineation parallel to the fold hinges.

Kinematic criteria at macro- and microstructural scale in the Tarroiba cliff do not indicate a preferred sense of movement: 1)the geometry of the quartz polycrystalline aggregates in the shear bands is similar to that of quartz polycrystalline aggregates within the pods 2) the ECCs are mainly symmetric within the shear bands, and 3) c-axis quartz fabrics have an orthorhombic symmetry. The existence of syntectonic flattened garnets with white mica enclosing them and thermobarometric data, suggest that formed the mylonitic foliation under*HT-HP* conditions.

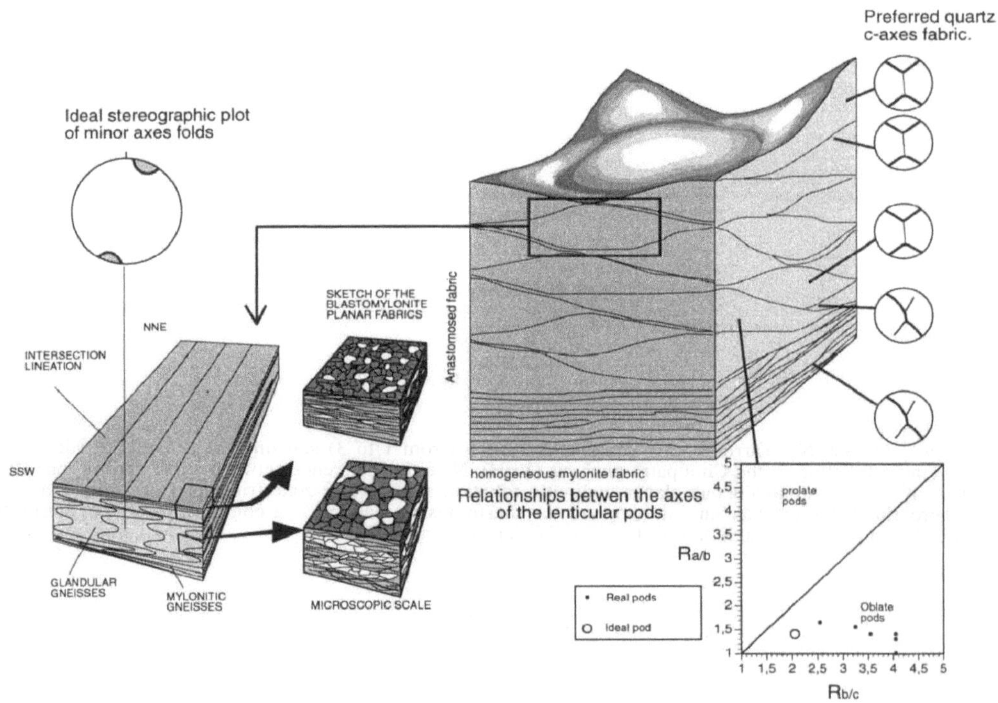

Figure 12. Conceptual model for the distribution of macro- and microstructural features in a mylonitic foliation developed by heterogeneous pure shear, based on data from the Chímparra gneisses.

Both macro- and microstructural analyses suggest heterogeneous pure shear or flattening as the main deformation process related to the development of the *HP-HT* mylonitic foliation prior to re-equilibration under amphibolite facies conditions (Fig.12). However, some microstructural characteristics are not easy to interpret, The quartz c-axis fabric displays a maximum which is probably due to basal <a> glide and which is characteristic of lower temperature than produced the garnet flattening. If the *HP-HT* mylonitic foliation is well-preserved, why is there no evedence of the *HP-HT* quartz c-axis fabrics? Or conversely why does low temperature c-axis fabric occur in rocks in which the *HP-HT* paragenesis is so well preserved? In addition, the absence of the microstructures commonly seen in simple shear zones and most of the field data suggest that the foliation is a pure shear feature. Therefore, the anastomosing shear bands relate to the heterogeneous distribution of the deformation, and according to the model proposed by Bell (1978, 1981), they can be interpreted as having formed by heterogeneous pure shear.

Acknowledgments

We thank Joaquina Alvarez-Marrón, Robert Ward and Dennis Brown for helpful discussions and improvement of the English version of the manuscript. This work was supported by the Spanish Dirección General de Investigacion Científica y Técnica (DGICYT) grant PB92 1022.

REFERENCES

Arenas, R., Gil Ibarguchi, J. I., Gonzalez Lodeiro, F., Klein, E., Martínez Catalán, J. R., Ortega Gironés, E., Pablo Maciá, J. G. and Peinado, M. (1986) Tectonostratigraphic units in the complexes with mafic and related rocks nothwest of the Iberian Massif, *Hercynica*, II, 2, 87-110.

Bastida, F. (1993) A new method for the geometrical classification of large data sets of folds, *J. Struc. Geol.*, Vol. 15, 1, 69-78.

Bell, T. H. (1978) Progressive deformation and reorientation of fold axes in a ductile mylonite zone: The Woodroofe thrust, *Tectonophysics*, 44, 285-320.

Bell, T. H. (1981) Foliation development. The contribution, geometry and significance of progressive, bulk, inhomogeneous shortening, *Tectonophysics*, 75, 273-296.

Blacic, J. D. (1975) Plastic mechanic deformation in the quartz the H_2O effect, *Tectonophysics*, 27, 271-294.

Boullier, A. M. and Bouchez, J. L. (1978) Le quartz en rubans dans les mylonites, *Bull. Soc. Geol. France*, 3, 253-262.

Bouchez, J. L. and Pecher, A. (1981) The Himalayan main central thrust pile and its quartz-rich tectonites in central Nepal, *Tectonophysics*, 78, 23-80.

Driessche, J. and Brun, J. P. (1987) Rolling structures at large shear strain, *J. Struct. Geol.*, 9, 5/6, 691-704.

Engels, J. P. (1972) The catazonal poly-metamorphic rocks of the Cabo Ortegal (NW Spain): a structural and petrofabric study. *Leidse Geol. Meded*, 48, 1-133.

Fernández, F. (1993) La deformación de los gneises de Chimparra en Punta Tarroiba (Cabo Ortegal, NW de España), *Rev. Soc. Geol. España*, 6(3-4), 77-91.

Fernandez, F. J. (in press) *Estructuras desarrolladas en gneises bajo condiciones de alta P y T, (Gneises de Chimparra, Cabo Ortegal)*, Lab. Xeoloxico de Laxe, Serie Nova Terra, Ediciós do Castro, Sada, A Coruña.

Gil Ibarguchi, J. I. and Arenas, R. (1990) Metamorphic evolution of the allochthonous complexes from northwest of the Iberian Peninsula, *In:* R. D. Dallmeyer and E. Martínez García (Eds.), *Pre-Mesozoic geology of Iberia*, 237-246, Springer-Verlag, Berlín.

Gil Ibarguchi, J. I., Mendia, M., Girardeau, J. and Peucat, J. J. (1990) Petrology of eclogites and clinopyroxene-garnet metabasites from the Cabo Ortegal Complex (northwestern Spain), *Lithos* , 25, 133-162.

Hanmer, S. and Passchier, C.W. (1991) Shear sense indicator: a review. *Geol. Surv. Can.* Pap. 90-17.

Hudleston, P. J. (1973) Fold morphology and some geometrical implications of theories of fold development, *Tectonophysics*, 16,1-46.

Ji, S. and Martignole J. (1994) Ductility of garnet as an indicator of extremely high temperature deformation, *J. Struct. Geol.*, 16, 7, 995-996.

Law, R. D. (1990) Crystallographic fabrics: a selective review of their applications to research in structural geology, *In:* R. J. Knipe y E. H. Rutter, *Deformation mechanism, rheology and tectonics,* Geol. Soc. Spec. Pub, 54, 335-353.

Lister, G. S. and Hobbs, B. E. (1980) The simulation of fabric development during plastic deformation and its application to quartzite: fabric transitions, *J. Struct. Geol.,* 2, 355-370.

Marcos, A., Marquínez, J., Pérez-Estaún, A., Pulgar, J. A. and Bastida, F. (1984) Nuevas aportaciones al conocimiento de la evolución tectonometamórfica del Complejo de Cabo Ortegal (NW de España), *Cuad. Lab. Xeol. Laxe,* 7, 125-137.

Nicolas, A. and Poirier, J. P. (1976) *Crystalline plasticity and solid-state flow in metamorphic rocks,* Wiley, New York, 444 pp.,

Pérez-Estaún, A., Martínez-Catalán, J. R. and Bastida, F. (1991) Crustal thickening and deformation sequence in the footwall to the suture of the Variscan belt of northwest Spain, *Tectonophysics,* 191, 243-253.

Prior, D.J. (1993).- Sub-critical fracture and associated retrogression of garnet during mylonitic deformation. *Contrib. Miner. Petrology, 113, 4, 545-556.*

Ramsay, J. G. (1967) *Folding and fracturing of rocks,* McGraw-Hill, New York and London. 568 pp.

Ries, M. y Shackleton, R. M. (1971) Catazonal complexes of northwest Spain and north Portugal, remnants of a Hercynian thrust plate, *Nature, Phys. Sci.,* 234, 65-68.

Tullis, J. A., Christie J. M. y Griggs D. T. (1973) Microstructures and preferred orientations of experimentally deformed quartzites,*Geol. Soc. Amer. Bull.,* 84, 297-314.

Vogel, D. E. (1967) Petrology of an eclogite- and pyrigarnite-bearing polymetamorphic rock complex at Cabo Ortegal, NW Spain, *Leidse Geol. Meded.,* 40, 123-213.

White, S. H. (1979) Paleostress estimates in the Moine thrust zone, *Nature,* 280, 222-223.

COUNTRY ROCK DISPLACEMENT DURING EMPLACEMENT OF THE JOSHUA FLAT PLUTON, WHITE-INYO MOUNTAINS, CALIFORNIA

ECKARDT STEIN
Institute for Mineralogy, TH Darmstadt
D-64287 Darmstadt

SCOTT R. PATERSON
Department of Earth Sciences, University of Southern California,
Los Angeles, CA 90089

1. Abstract

The Jurassic Inyo batholith is a 70 by 25 km, rectangular-shaped, NW-SE oriented body consisting of several distinct internal plutons, which range from monzonite to granodiorite. This batholith intruded upper Proterozoic to lower Paleozoic anchizonal to greenschist facies, multiply deformed metasediments during which a narrow (.004 to .16 batholith radii) structural aureole formed in the country rock. The structural characteristics of this aureole vary from rather discordant margins to regions dominated by complex ductile deformation.

The southern end of the batholith consists of the zoned Joshua Flat pluton with an outer monzonite displaying widespread cm- to dm-scale, margin-parallel compositional magmatic banding and magmatic foliations with foliation intensity and presence of this magmatic banding decreasing toward the center of the pluton. The magmatic foliation crosscuts compositional banding and thus formed later. A complex, but narrow, structural aureole is present in this region. A relatively abrupt transition occurs at the outer margin of the aureole where folded, but relatively flat-lying lower to upper Cambrian metasediments become steeply NE-dipping. Progressing inwards, bedding and bedding parallel cleavage show a fan-like, but largely inward-dipping (=dipping toward the pluton contact) structure. Stratigraphic units everywhere face inwards and are locally overturned near the batholith, an observation that has been used by others to argue for country rock return flow during emplacement. However, in the inner aureole, chocolate tablet boudins in competent layers, indicating thinning by ductile flattening, are associated with discrete ductile mylonite zones, asymmetric folds, and moderately plunging mineral lineations which support oblique sinistral shear in a SW domain and dextral shear in a SE domain.

Initial results point to a combination of downward displacement and oblique SE and SW ductile flow occurring during batholith emplacement. Our initial estimates of bulk shortening in the aureole indicate that at least 40% to 60% of the aureole is missing. Well preserved and undeflected regional markers outside the aureole indicate that the additional transfer of country rock is needed during emplacement and must involve a large amount of vertical displacement within the region now occupied by the batholith. Local discordant margins and rare xenoliths in the magma indicate that the process of stoping probably accounts for some of the missing aureole.

O. Oncken and C. Janssen (eds.), Basement Tectonics 11, 35–49.

2. Introduction

For over 200 years geologists have attempted to understand how magma is emplaced in the crust, a non-trivial problem given the large volumes of plutonic material in many orogenic belts. Paterson and Fowler (1993) noted that there are only four "space making" mechanisms during pluton emplacement in the crust: (1) lowering the Moho, (2) displacement of the Earth's surface, (3) volume loss, and (4) elastic contraction. All other pluton emplacement mechanisms are better viewed as material transfer processes (MTPs) since they move material within the crust, but do not change the volume (or density) of the crust. If magma generation occurs in the crust, magma ascent and emplacement require only crustal-scale MTPs. These authors also argued that because of the highly variable and rapidly changing conditions in magma-country rock systems, MTPs will likely vary with time, depth, and distance from the pluton (e.g., Buddington, 1959). Below we examine structures in the aureole of the SE edge of the Inyo batholith (Figs. 3, 4), in the White-Inyo Mountains and evaluate likely emplacement MTPs.

Figure 1. Geographic overview of the southwestern most part of the USA with the position of the White-Inyo Mountains (black rectangle).

The White-Inyo Mountains are situated in the east part of central California (Fig. 1). It is part of the westernmost Basin and Range province (Nelson, 1981) which is characterized by a continuous approximately N-S-oriented succession of mountain ranges and narrow basins. The White-Inyo Mountains form a horst separated from the Sierra Nevada to the west by the Owens valley and to the east by the Quarternary Fish Lake, Eureka and Saline valley basins (Fig. 2).

The oldest exposed units in the White Mountains are strata of the Cordilleran miogeocline of Precambrian to early Paleozoic Inyo succession (Nelson, 1962; Stewart, 1970, 1972). The basement is unknown, but a continental affinity is indicated by the Sr isotopic signature of Mesozoic plutons in the region (Kistler and Peterman, 1973). Middle and late Paleozoic strata overlie the Inyo succession and along with the older strata have been metamorphosed under greenschist grade conditions (e.g. Ernst et al., 1991). Many retain visible sedimentary fabrics. Mesozoic volcanic rocks and their volcaniclastic counterparts are present immediately north of the study area (Krauskopf, 1971; Crowder and Sheridan, 1972; Hanson, 1986; Ernst et al., 1991).

At least three periods of regional folding and thrusting have been documented in the White-Inyo Mountains which are thought to range in age from Permian to late Jurassic (Stewart et al., 1966; Nelson, 1971; Sylvester and Babcock, 1975; Corbett et al., 1988; Snow, 1992). Although there is some controversy about the age of these structures, previous workers generally agree that all the thrusts and large-scale folding predate emplacement of the Jurassic Inyo batholith (Nelson, 1980). Thus these older structures and stratigraphic units provide excellent regional markers that can be used to examine pluton emplacement processes (e.g., Fowler, 1994).

Deformation of Cretaceous age has been documented in the White and Inyo Mountains (Dunne et al., 1978; Hanson, 1986, 1987; Hanson et al., 1987) but has not be recognized near the Inyo batholith. Cenozoic extensional faulting has caused uplift and some tilting of the whole range and formation of local basins (Dunne et al., 1978; Nelson, 1980).

3. Inyo Batholith

The Inyo batholith forms part of a Jura-Triassic, locally alkali-rich arc extending at least from British Columbia to Mexico (Miller, 1978). This batholith is an approximately 70 by 25 km, rectangular-shaped, NW-SE oriented body consisting of several different phases (Fig. 3): the Eureka Valley monzonite, the Marble Canyon monzodiorite, the Joshua Flat monzonite, and the Beer Creek granodiorite. Isotopic potassium-argon data from hornblendes and biotites of these rocks (Kistler et al., 1965; McKee & Nash, 1967; Evernden & Kistler, 1970) show 2 distinct time spans from 179-168 Ma (hbl) and from 170-162 Ma (bio). U/Pb dates for zircons from the Beer Creek granodiorite and the Joshua Flat monzonite are in the same 160-180 Ma range (Sylvester et al., 1978a; Gillespie, 1979). Most of the batholith is continuous: however, the satellite Marble Canyon pluton occurs to the SE (Fig. 3). The northern end of the batholith is locally intruded by Cretaceous plutons.

This batholith is petrologically interesting in that it is a quartz-poor, high K_2O suite with abundant incompatible trace elements (Miller, 1978; Sylvester et al., 1978b). Sr ratios in the Joshua Flat pluton are somewhat lower (0.705 to 0.706) than normal values (0.706 to 0.709) for plutons in the White-Inyo Mountains (Sylvester et al., 1978b). Sylvester et al. (1978a) argued that these magmas are not petrogenetically related to adjacent Sierra Nevada batholith, but instead were some of the first formed,

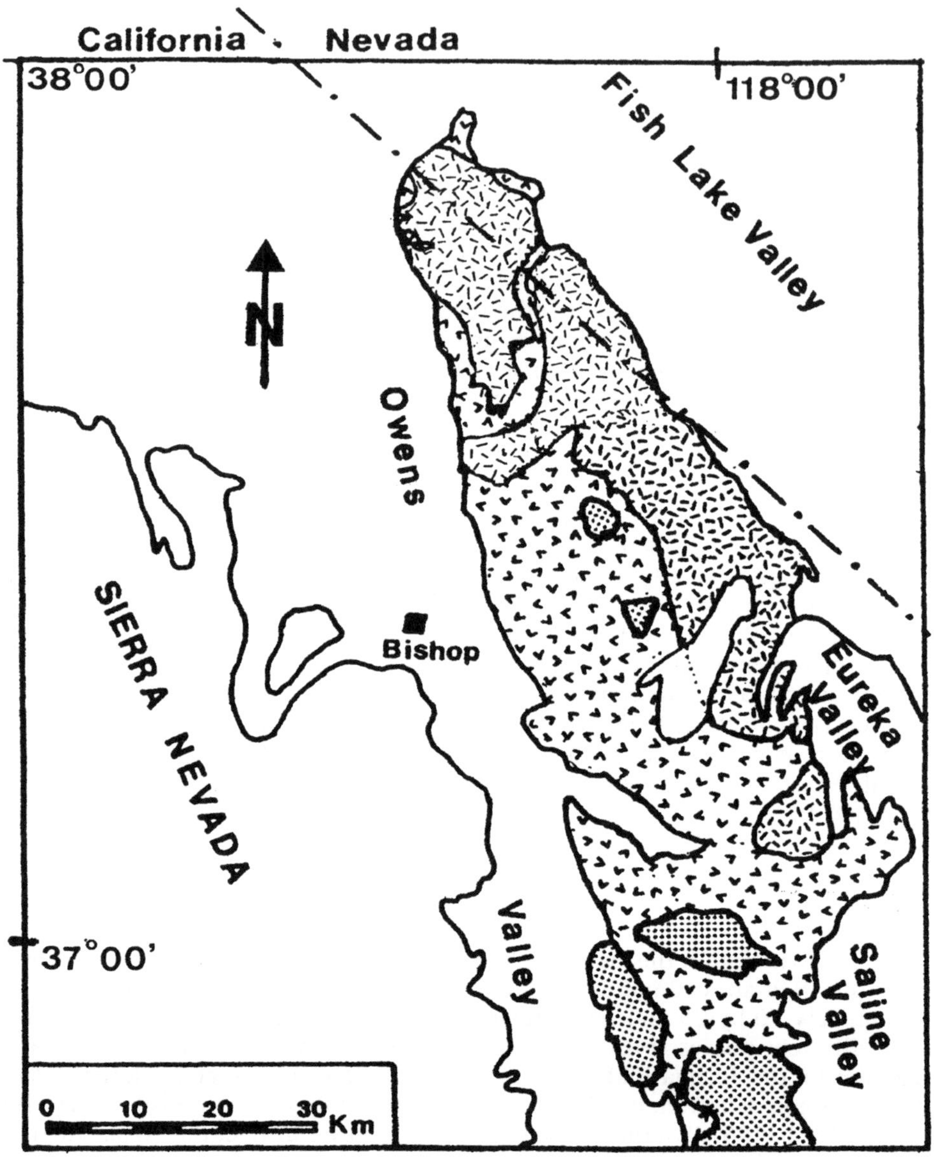

Inyo Batholith

undivided plutonic rocks in the White-Inyo Mountains

undivided metasediments and volcanics in the White-Inyo Mountains

quarternary sediments in the White-Inyo Mountains

Figure 2. Schematic geological map of the White-Inyo Mountains (modified after Sylvester et al. 1978) easternmost, and deepest magmas in this arc and were derived from partial melts of upper mantle.

At the southern end of the batholith, the Joshua Flat pluton (Figs. 3, 4) is characterized as a medium grained hornblende - augite monzonite with fine grained quartz monzonitic, granodioritic and dioritic to quartz dioritic intercalations (Nelson, 1966a). Below we focus on work in progress at the southern end of the Inyo batholith, that is remapping of the Joshua Flat pluton and the SW, S and SE parts of its contact aureole at a scale of 1:10,000 (Fig. 4). The northern part of the aureole is not visible because here the pluton is intruded by the Beer Creek phase of the Inyo batholith.

4. Stratigraphy in Aureoles

Nelson (1966a, 1966b, 1971) recognized 6 lower to upper Cambrian formations in the aureole of the Joshua Flat pluton, which he interpreted as a succession of clastic and carbonate passive margin shelf sediments.

Lead Gulch Formation (*upper Cambrian*): brown, siliceous hornfels; cherts; gray, banded marble; and calcsilicate-hornfels.

Bonanza King Dolomite (*middle to upper Cambrian*): fine-laminated to thick-bedded, fine grained, light to dark gray dolomite; and coarse grained, white calcite-marble.

Monola Formation (*middle Cambrian*): siliceous hornfels and calcsilicate-hornfels interbedded with fine grained limestone and siltstone.

Mule Spring Limestone (*lower Cambrian*): laminated, fine grained to thick-bedded, medium grained, gray limestone and marble.

Saline Valley Formation (*lower Cambrian*): siliceous hornfels; thin-bedded siltstones; fine grained quartzite; gray silty shale; fine grained, thin-bedded limestone and massive quartzite.

Harkless Formation (*lower Cambrian*): gray siltstone; platy to thick-bedded, gray, fine to medium grained quartzite, and minor limestone intercalations.

The Saline Valley Formation can be subdivided into an upper member consisting of shales, siltstones, marls, and limestones, and in a lower member consisting of thick, homogeneous sandstones. Crossbedding as well as ripple marks in the metapsammitic rocks of the upper member of the Saline Valley Formation provide clear indicators for younging directions, fold vergence, and sedimentary transport directions.

5. Structures

5.1. Joshua Flat Pluton

The Joshua Flat pluton is composed of three internal phases consisting of monzonite to layered granodioritic-monzonitic rocks, which form a magmatic layering in 100 m to km scale. A cm- to dm-scale, steeply inclined, compositional magmatic banding ranging from monzonitic, granodioritic to granitic lithologies occurs in these units and is oriented parallel to the batholith margin. This banding is enhanced by pronounced grain size differences. A roughly margin parallel, steeply inclined, magmatic foliation defined by aligned biotite and hornblende is well developed in this pluton. Locally this foliation crosscuts the magmatic banding, indicating that the magmatic foliation developed later.

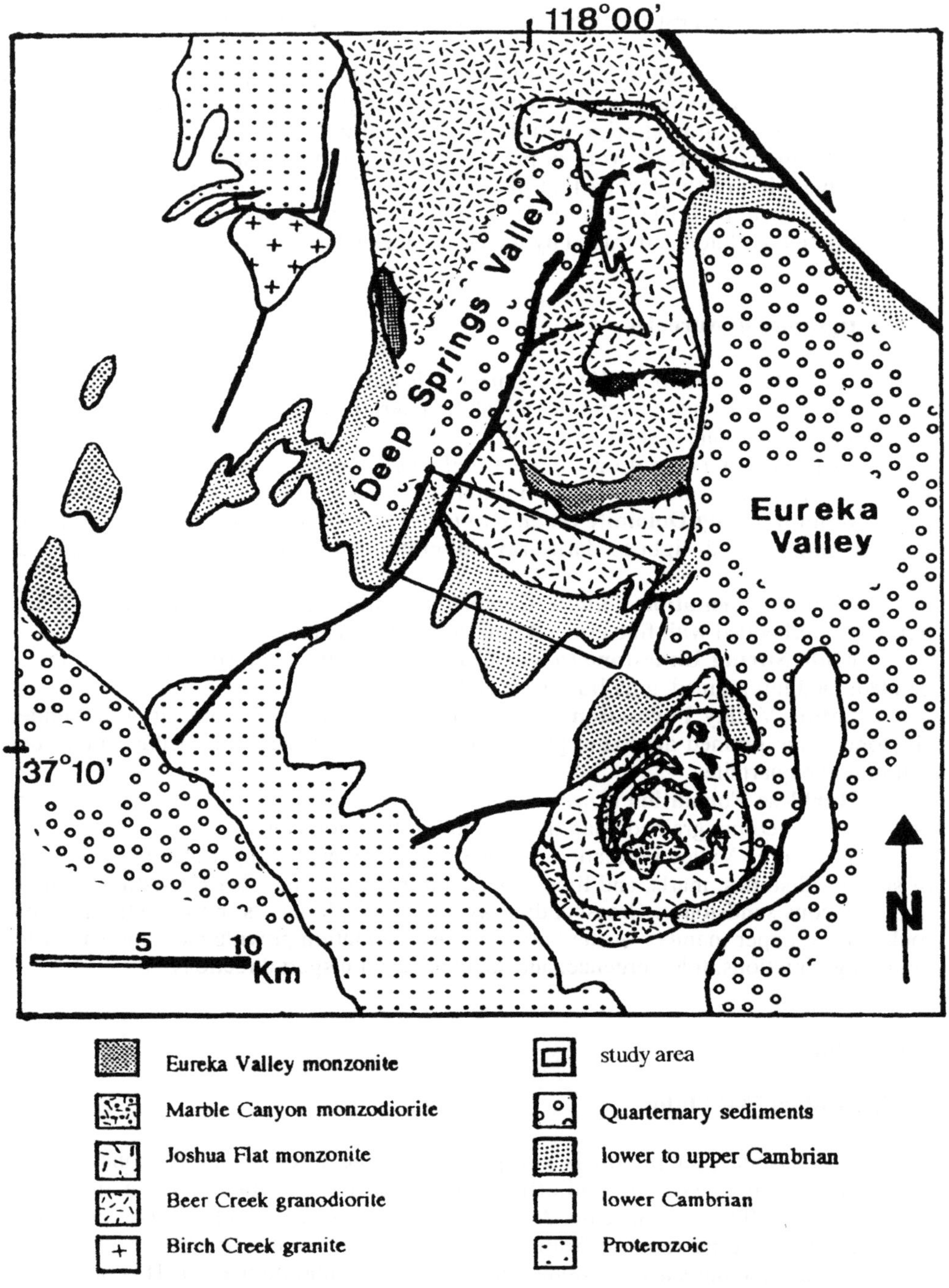

Figure 3. Schematic geological map of the S part of the Inyo Batholith (modified after Nelson, 1981)

Strongly flattened, fine grained, mafic enclaves with dioritic compositions are abundant along the southern margin, and their long axes are oriented parallel to the magmatic banding. Sometimes the magmatic foliation wraps around the pancake-shaped enclaves, indicating that the shapes developed early in the magmatic history. The monzonite also contains several different sized (1-100 m) lenses of hornfelsic xenoliths representing stoped blocks of the country rocks. The xenoliths do not show any preferred orientations, or syn-magmatic strain, which emphasizes their different rheological behaviour when compared to the enclaves.

Toward the pluton center, the monzonite becomes more homogeneous, magmatic banding and enclaves are less common and less flattened, and only rare magmatic layering of mafic-richer and mafic-poorer rocks can be observed. Magmatic minerals show only a weak alignment. No penetrative solid-state deformation has as yet been observed in this part of the Joshua Flat pluton, therefore all the planar fabric elements described above are interpreted to be magmatic in origin.

There are other small magmatic bodies occurring in the Joshua Flat pluton as well as in the aureole such as porphyritic dikes, diorites and granites. Based on their fabrics, related skarns, narrow contact aureoles, and evidence of stoping, they are interpreted to be younger plutons intruded into a colder and shallower environment.

5.2. Contact Aureole

Everywhere around the SW, S and SE end of the Inyo batholith, a narrow structural aureole in the lower to upper Cambrian formations is developed in which regional structures gradually steepen toward the pluton margin and locally right next to the contact are overturned. Our examination of the aureole at the southern end of the batholith (Fig. 4) shows the following. Outside the aureole upper Proterozoic to lower Cambrian strata are upright and, along with a bedding parallel cleavage (s_1), are gently to moderately ESE-dipping. This orientation is due to a F_2-folding event which developed open, upright to E vergent, km-scale folds with horizontal B_2-fold axes showing a NE-SW strike direction. The area S of the Joshua Flat pluton is situated on the eastern flank of such an anticline. A related crenulation with discrete cleavage planes (s_2) is restricted to metapelitic rocks. A B_2-fold axes parallel intersection lineation ($ss//s_1>s_2$) is the only linear fabric in the regional metamorphic rocks. The regional metamorphic imprint outside the contact aureole is in lower greenschist facies, characterized by the prograde mineral assemblage sericite/muscovite and chlorite. According to Ernst et al. (1993) peak metamorphic conditions reached 375±75°C and 3 to 4 kbars.

A relatively abrupt structural transition occurs at the outer margin of the aureole. Generally the contact aureole´s rocks consist of a younger, lower to upper Cambrian sequence. This aureole sequence becomes steeply NE-dipping. Progressing inwards (i.e., into younger stratigraphic units), planar fabrics like bedding as well as bedding parallel cleavage show a fan-like structure when viewed in a N-S cross section (Fig. 5), with units immediately adjacent to the contact dipping steeply SSW. Crossbedding and ripple marks in quartzites of the lower Cambrian upper Saline Valley Formation clearly indicate that the SSW-dipping planes near the batholith are upside down. All of the contact aureole rocks are characterized by a pronounced planar fabric defined by the parallelism of irregular spaced sedimentary bedding (ss) and a mm-spaced foliation (s_1). This foliation corresponds to the regional s_1 outside

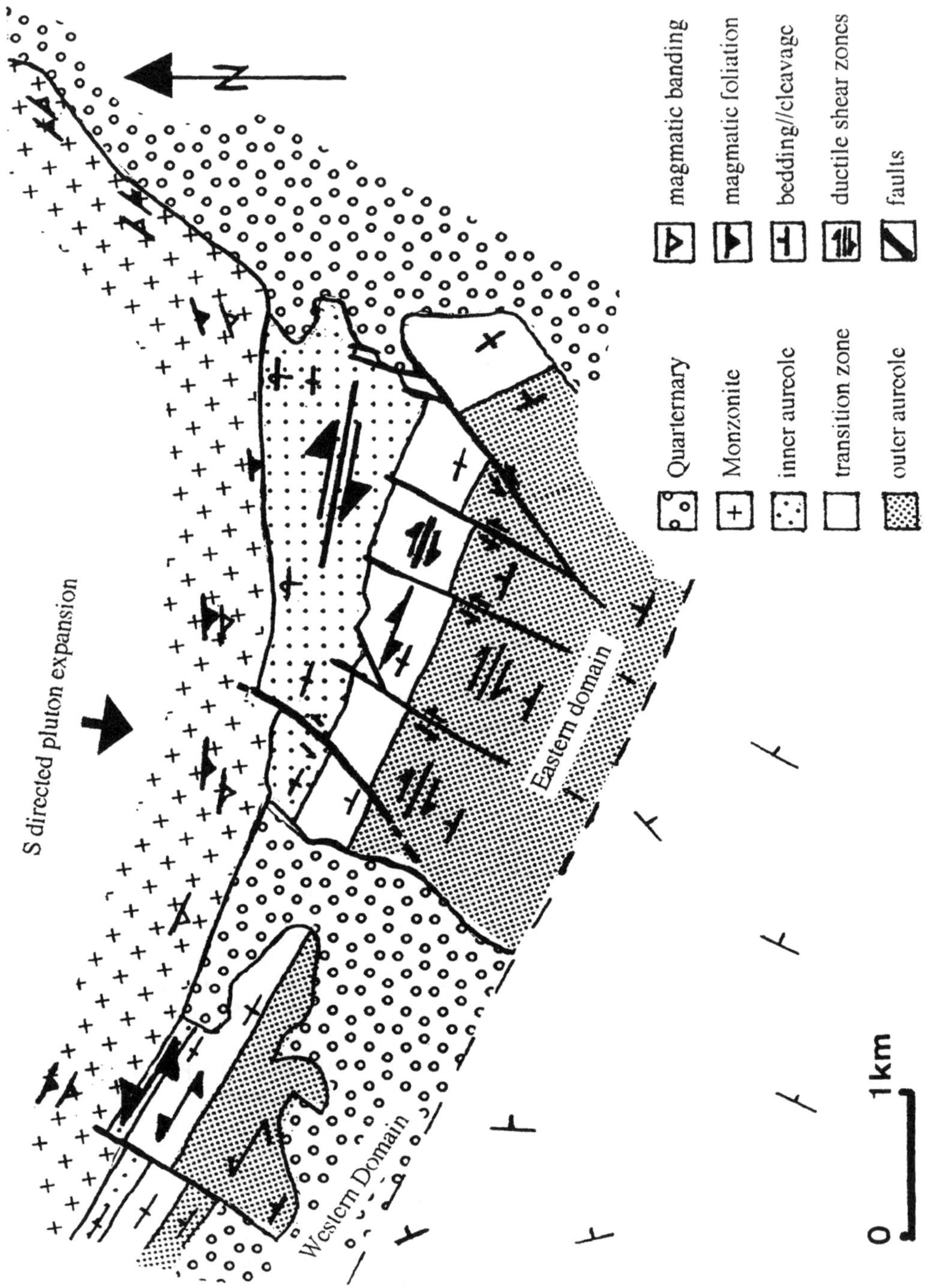

Figure 4. Schematic geological map of the S part of theJoshua Flat pluton

the aureole but inside the structural aureole it was reactivated and overprinted by deformations due to the emplacement of the plutons.

In the outer structural aureole (Figs. 4, 5), represented by the upper parts of the Harkless Formation, lower and upper Saline Valley Formation, and part of the Mule Spring Limestone, ductile fabrics caused by flattening and oblique shear occur only in discrete zones along lithologic boundaries with prominent changes in rheology and/or along the formation boundaries. Stratigraphic units maintain a relatively constant thickness, although bedding-parallel strike-slip as well as dip slip during ductile flattening and oblique slip during ductile shear is common. Syn-emplacement, discontinuous faults with a horizontal offset of about 50 to 100 m are abundant in the outer aureole. A mineral assemblage containing andalusite, biotite and muscovite indicates upper greenschist facies conditions during contact metamorphism. Microstructures like static mineral growth, equilibrated grain boundaries with triple points especially in quartz and feldspar fabric and the lack of plastic crystal lattice deformations even in quartz indicate that metamorphism outlasted deformation.

The inner aureole (Figs. 4, 5) represented by the Lead Gulch Formation, the Bonanza King Dolomite and partly by the Monola Formation shows a syn- to postkinematic contact metamorphism in pyroxene-hornfels facies to amphibolite facies. This succession is characterized by strong lateral changes in stratigraphic thickness, interpreted as the result of an intense ductile deformation, and by the absence of emplacement related brittle faults. Well developed bedding parallel cleavage (s_{2a}), chocolate tablet boudins with short horizontal axes perpendicular to the steeply inclined to subvertical bedding, and long axes both subhorizontal and subvertical indicate thinning of bedding by ductile flattening. Discrete ductile mylonite zones with open to close, noncylindrical, disharmonic, intrafolial folds from cm- to dekameter scale become increasingly abundant toward the pluton contact. An axial-planar crenulation cleavage (s_{2b}) is restricted to the hinge zones and short limbs of these folds. Strong boudinage developed only in the long fold limbs, indicating that flattening and ductile shear occurred contemperaneously. Mineral lineations and kinematic indicators, such as sc-fabrics and σ- and δ-clasts define two different structural domains in the mapped aureole, a western domain and an eastern domain (Fig. 6). The western domain is characterized by an oblique sinistral shear, and frequent mineral lineations and mineral stretching lineations parallel to the fold axes. In contrast the SE domain is characterized by an oblique dextral shear, and scarce mineral lineations and mineral stretching lineations at a large angle to the subvertical fold axes. Diverging fold vergences to the SW and to the SE in the different structural domains support the existence of these two domains.

In both domains, units from Lead Gulch to Mule Spring Limestone contain close to tight, recumbent folds together with a related widely spaced subhorizontal fracture cleavage (s_3) in areas with steeply inclined fabrics, that overprint all of the above mentioned ductile structures. Fold axis orientation is controlled by the strike direction of the steeply inclined planar fabrics and fold vergence by the dip direction, for example, NE-dipping planes develop NE facing folds, and SW-dipping planes develop SW facing folds.

Our initial mapping results suggest, that the extent of the inner and outer aureole is largely a function of distance from the pluton and is not controlled by rock type, and thus predominantly reflects heat from the pluton. However, there is a

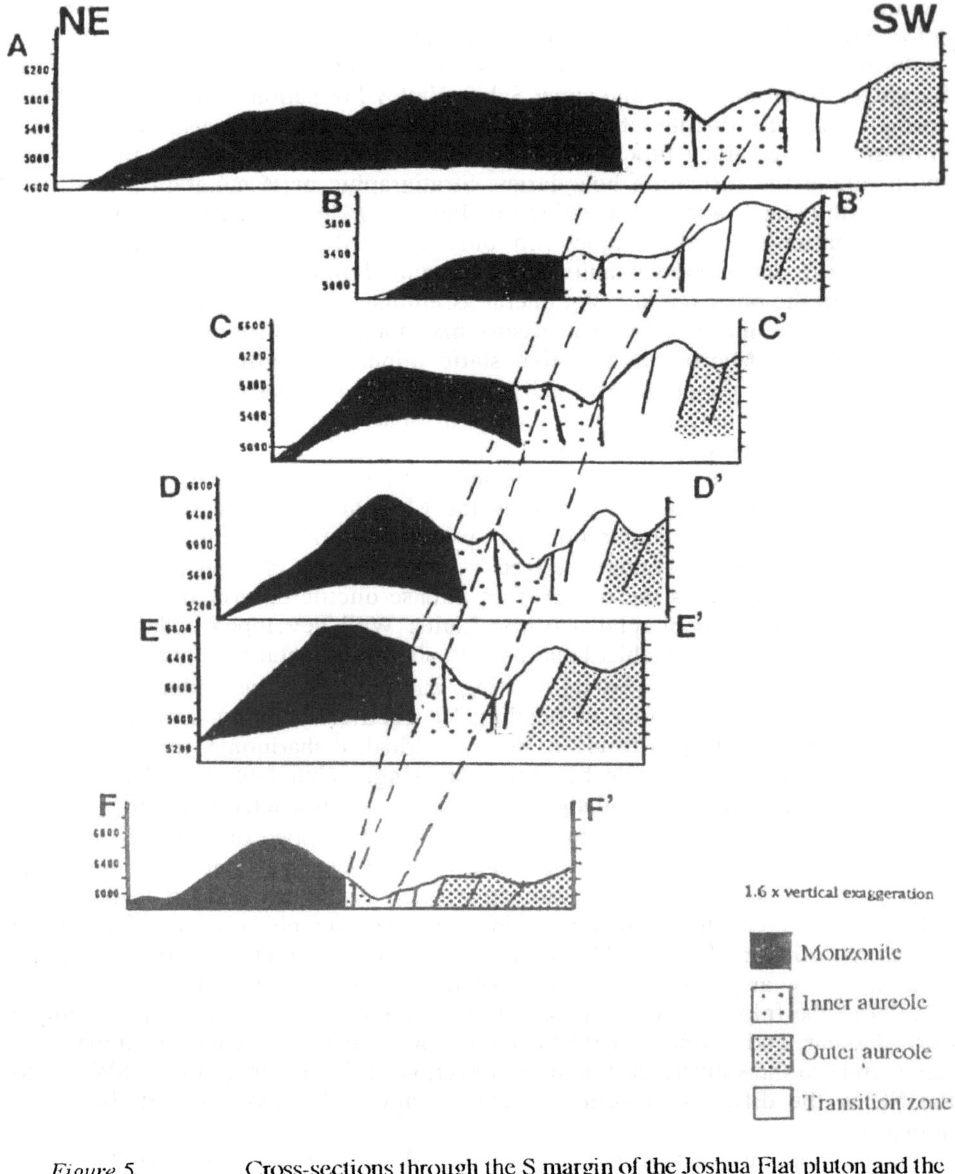

Figure 5. Cross-sections through the S margin of the Joshua Flat pluton and the E domain of its aureole

transition zone where stratigraphic control on rheology becomes more important. This zone occurs in parts of the Monola Formation and Mule Spring Limestone, and is characterized by both ductile and brittle deformations which are related to the pluton emplacement. Strain responsible for lateral ductile flow indicated by mylonite zones and dishamonic, internal folds occur only along discrete zones at formation boundaries or along other important changes in rock type. Ductile folding is replaced by single or conjugate kink bands. Despite its greater distance from the pluton in the E domain, the Mule Spring Limestone of the E domain shows more ductile deformation than the adjacent, more siliceous Monola Formation. In the W domain, where it is closer to the pluton, the Monola Formation also shows flattening and thinning by ductile strain.

Summary of the Western and Eastern Structural Domain

Folds:

(W) mostly rootless, **gently plunging, W-vergent**
(E) mostly rootless, **steeply plunging, E-vergent**
(W + E) tight, recumbant folds

S-surfaces: no differences in both domains

sedimentary bedding (ss),
s_1, bedding parallel, regional, reactivated during emplacement
s_{2a}, margin parallel boudinage in long fold limbs,
s_{2b} crenulation in fold hinges,
s_3 subhorizontal fracture cleavage, restricted to subvertical areas

Boudinage: no differences in both domains

chocolate tablet boudins in long limbs of folds, flattening

Shear zones:

(W) steep, ± bedding or formation boundary parallel, **sinistral**
(E) steep, ± bedding or formation boundary parallel, **dextral**

Lineations:

(W) mineral, stretching, **frequent** in Lead Gulch Formation, Bonanza King Dolomite, and Monola Formation, **parallel to fold axes 240/30** *
intersection lineation ($ss//s_1:s_2$) parallel to fold axes.
(E) mineral, stretching, **scarce, 120/50** *, intersection lineation ($ss//s_1:s_2$) parallel to fold axes 120/80-260/80 *

(* examples from the inner aueole)

Fig. 6 Characteristic syn-emplacement fabric elements of the two structural domains

Initial studies along the E and W margins of this pluton indicate that the aureole remains narrow along the sides, and that ductile flow is generally toward the SW on both sides, which means toward the "corners" of the batholith, indicating complex flow and huge space problems at these corners (Stein, 1994), because aureole material is flowing toward the SW- and SE-corner from both the end and sides. These field relationships give rise for the interpretation of a forceful expansion of the Joshua Flat pluton during later stages.

Brittle faults that crosscut the pluton and the inner and outer aureole are interpreted to postdate pluton emplacement and at least partly reactivate older syn-emplacement structures.

6. Emplacement Mechanisms

Supported by the results from the S edge of the Joshua Flat pluton, downward return flow of country rock is interpreted as the earliest, and one of the more important material transfer mechanisms during emplacement of the Joshua Flat pluton. The inward-bending of formations within the structural aureole all around the pluton and the systematic decreasing age of the stratigraphy within the aureole toward the pluton clearly document an increasing amount of downward movement of country rock as the pluton is approached. Stratigraphic units in the aureoles of both the Joshua Flat and Marble Canyon plutons generelly are younger (lower to upper Cambrian) than the adjacent country rocks outside the aureoles (Proterozoic to lower Cambrian) and sometimes topographically lower as well. Early steepening and inward-bending of stratigraphic units is also required by the attitudes of stratigraphic units in the SW and SE domains given the gentle ESE-dips of units immediately outside the aureole. Our initial observations indicate that downward flow occurred by bedding-parallel slip and some localized crystal-plastic flow.

Structures associated with flattening and oblique lateral shear overprint the structures caused by downward flow and were more likely formed due to a S-directed movement (due to either in situ expansion or continued rise of an elliptical body) of the pluton, which deformed existing steeply inclined strata rather than the gently ESE-dipping units, because an earlier expansion would have caused SW-directed overthrusts, and SW vergent folds in gently ESE-dipping strata. This flattening resulted from pervasive crystal-plastic flow during at least amphibolite facies metamorphism in the inner aureole, and bedding-parallel slip and brittle faults in the outer aureole during lower greenschist facies metamorphism.

The late recumbant folds in the inner aureole indicate that some vertical shortening occurred either late in the emplacement process or after emplacement was complete. But whether this shortening was due to gravitational forces or to the latest pulses of magma emplacement remains unclear.

Stoping is indicated by stoped blocks throughout the Marble Canyon pluton, in the SE part of the Joshua Flat pluton and by discordant irregular margins in the NW part of the Inyo batholith. Rock types of stoped blocks in the Joshua Flat pluton do not match the aureole rocks, therefore timing, importance and spatial relationships remain unclear. The fact that the stoped blocks have only a weak metamorphic banding, without any flattening fabric, may indicate that at least some stoping occurred relativly early in the emplacement history.

Assimilation of crustal material during ascent and emplacement of the magma is of minor importance, because geochemical data like Sr initials of about 0.705 to 0.706 point to melts derived from the upper mantle (Sylvester et al., 1978b).

7. Summary and Conclusions

Length considerations do not allow us to fully explore and defend an emplacement history for the Joshua Flat pluton and Inyo batholith. However, the following reflects some initial speculations on this subject. Units within the Joshua Flat pluton show a crude normal zoning, compositionally and texturally distinct, magmatic layering, a narrow compositional magmatic banding which is abundant at the pluton's margin, and a roughly margin-parallel, magmatic foliation, which overprints the banding. Our initial interpretation of these igneous features is that three somewhat distinct magma pulses arrived in the chamber with younger pulses intruding the inner, less viscous part of the chamber. Gradual compositonal changes within each pulse reflect fractionation, whereas the strong magmatic banding, which always occurs along contacts, reflects simultaneous fractionation and contact parallel flow within a steep velocity gradient. Magmatic foliation formed during late strain of the melt and reflects the regional plus emplacement-related strain field during final crystallization.

Structures in the southern aureole include (1) pre-emplacement bedding parallel foliation, (2) syn-emplacement fanning structure formed by reorientation and steepening of Cambrian formations during downward flow and ductile high-temperature foliation, boudinage, mylonite zones, and intrafolial folds in an inner aureole, and a fracture cleavage and brittle faults in an outer aureole, and (3) post-emplacement recumbent folds and brittle faults that crosscut the aureole and pluton.

These structures define a narrow contact aureole (about 0.004 to 0.16 pluton radii), in which estimates of bulk shortening (calculated from change in thickness of stratigraphic units) are around 50%. However, simplistic calculations of space made by retrodeforming this aureole suggest that only about 30% to a maximum of 60% of the space needed for emplacement can be accounted for by the deformation in this aureole. Therefore, at least 40% and possibly much more of the aureole is now missing. Pre-emplacement regional markers outside of this aureole show no emplacement-related deflections (Fig. 2), thus indicating that vertical MTPs operating in the region now occupied by the batholith must account for the remainder of the needed space.

Therefore we conclude that during the emplacement of the Joshua Flat pluton several different material transport mechanisms (stoping, downward flow, oblique crystal-plastic flattening, brittle faulting) were active but varied in importance through time and with distance from the pluton. Downward directed flow and possible a significant amount of stoping (to explain the missing aureole) are the oldest observable transport mechanisms and probably were the dominent MTPs during emplacement of the oldest magma pulse. Later pulses may have been emplaced by displacement of earlier magma and by oblique lateral ductile flow and ductile flattening in the inner aureole and brittle faulting in the outer aureole. This geometry of country rock flow creates severe space problems at the batholith's corners, regions that we have not yet remapped but that existing maps show are structurally complex. Finally, we note that this study reflects only a reexamination of one part of the aureole around the whole batholith.

8. References

Buddington, A.F. (1959) Granite emplacement with special reference to North America, *Geological Society of America Bulletin*, **70,** 671-747.

Corbett, K.P., Wrucke, C.T., and Nelson, C.A. (1988) Structure and history of the Last Chance thrust system, Inyo Mountains and Last Chance Range, California, *in* D.L. Weide and M.L. Faber (eds.), *This extended land, geological journeys in the southern Basin and Range,* Geological Society of America, Cordilleran Section, field trip guidebook, 269-292.

Crowder, D.F. and Sheridan, M.F. (1972) Geologic map of the White Mountain Peak quadrangle, Mono County, California, *U.S. Geological Survey Geologic Quadrangle Map* **GQ-1012**, scale 1: 62.500.

Dunne, G.C., Gulliver, R.M.,and Sylvester, A.G. (1978) Mesozoic evolution of the rocks in the White, Inyo, Argus, and Slate ranges, eastern Califonia, *in* D.G. Howell and K.A. McDougal (eds.), *Mesozoic paleogeography of the western United States*, Society of Economic Paleontologists and Mineralogists, Pacific Section, Pacific Coast Paleogeography Symposium **2**, 189-207.

Ernst, W.G., Hall, C.A.,Jr. and Nelson, C.A. (1991) Geologic map of the central White-Inyo Range, easternmost California, *Geological Society of America Abstracts with Programs,* **23/5,** 137.

Ernst, W.G., Nelson, C.A. and Hall, C.A. (1993) Geology and Metamorphic Mineral Assemblages of Precambrian and Cambrian Rocks of the Central White-Inyo Range, Eastern California, Central White-Inyo Range, Eastern California, Map Sheet 46, Division of Mines and Geology, Sacramento.

Evernden, J.F. and Kistler, R.W. (1970) Chronology of emplacement of Mesozoic batholithic complexes in California and western Nevada, *U.S. Geological Survey Professional Paper* **623,** 42

Fowler, T.K. Jr. (1994) Using geological maps to constrain pluton emplacement mechanisms, *Geological Society of America Abstracts with Programs*, **26**, 52.

Gillespie, J.G. (1979) U-Pb and Pb-Pb ages of primary and detritical zircons from the White Mountains, eastern California, *Geological Society of America Abstracts with Programs,* **11,** 79.

Hanson, R.B. (1986) Geology of Mesozoic metavolcanic and metasedimentary rocks, northern White Mountains, California, *[Ph.D. Thesis], University of California Los Angeles,* 231p.

Hanson, R.B. (1987) Superimposed deformations in the development of a shear zone: An example from the northern White Mountains, California, *Geological Society of America Abstracts with Programs,* **18/2,** 113.

Hanson, R.B., Saleeby, J.B., and Fates, D.G. (1987) Age and tectonic setting of Mesozoic metavolcanic and metasedimentary rocks, northern White Mountains, California, *Geology,* **15,** 1074-1078.

Kistler, R.W.; Bateman, P.C. and Brannock, W.W. (1965) Isotopic ages of minerals from granitic rocks of the central Sierra Nevada and Inyo Mountains, California, *Geological Society of America Bulletin*, **76/2,** 155-164

Kistler, R.W. and Peterman, Z.E. (1973) Reconstruction of crustal blocks of California on the basis of initial strontium isotopic compositions of Mesozoic granitic rocks, *U.S. Geological Survey Professional Paper* **1071,** 17.

Krauskopf, K.B. (1971) Geologic map of the Mount Barcroft quadrangle, California-Nevada: *U.S. Geological Survey Geologic Quadrangle Map* **GQ-960**, scale 1:62.500.

McKee, E.H. and Nash, D.B. (1967) Potassium-argon ages of granitic rocks in the Inyo batholith, east-central California, *Geological Society of America Bulletin*, **78**, 669-680.

Miller, C.F. (1978) An early Mesozoic alkalic magmatic belt in western North America, *in*: D.G. Howell and K.A. McDougal (eds.), *Mesozoic paleogeography of the western United States,* Society of Economic Paleontologists and Mineralogists, Pacific Section, Pacific Coast Paleogeography Symposium, **2**, 163-174.

Nelson, C.A. (1962) Lower Cambrian-Precambrian succession, White-Inyo Mountians, California, *Geological Society of America Bulletin,* **73,** 139-144.

Nelson, C.A. (1966a) Geologic map of the Waucoba Mountain quadrangle, Inyo County, California, *U.S. Geological Survey Geologic Quadrangle Map* **GQ-528**, scale 1:62.500.

Nelson, C.A. (1966b) Geologic map of the Blanco Mountain quadrangle, Inyo and Mono counties, California, *U.S. Geological Survey Geologic Quadrangle Map* **GQ-529**, scale 1: 62.500.

Nelson, C.A. (1971) Geologic map of the Waucoba Spring quadrangle, Inyo County, California, *U.S. Geological Survey Geologic Quadrangle Map* **GQ-921**, scale 1:62.500.

Nelson, C.A. (1980) Guidebook to the geology of a portion of the eastern Sierra Nevada, Owens Valley, and White-Inyo Range, *University of California Los Angeles, Department of Earth and Space Sciences Guidebook* **12**, 1-107.

Nelson, C.A. (1981) Basin and Range Province, *in* W.G. Ernst (ed.), *The Geotectonic Development of California*, **Rubey Volume I,** 202-216, Prentice-Hall, Englewood Cliffs, N.J.

Paterson, S.R. and Fowler, T.K. Jr. (1993) Re-examining pluton emplacement processes, *Journal of Structural Geology*, **15**, 191-206

Snow, J.K. (1992) Large magnitude Permian shortening and continental margin tectonics in the southern Cordillera, *Geological Society of America Bulletin* **104**, 80-105.

Stein, E. (1994) Structures in the aureole of the Joshua Flat Pluton: Implications for country rock flow patterns during emplacement, *Geological Society of America Abstracts with Programs, Cordilleran Section,* **26** , 95.

Stewart, J.H. (1970) Upper Precambrian ands Lower Cambrian strata in the southern Great Basin, California and Nevada, *U.S. Geological Survey Professional Paper* **620**, 1-206.

Stewart, J.H. (1972) Initial deposits of the Cordilleran geosyncline: evidence of a late Precambrian (<850 m.y.) continental separation, *Geological Society of America Bulletin*, **83,** 1345-1360.

Stewart, J.H., Ross, D.C., Nelson, C.A. and Burchfiel, B.C. (1966) Last chance thrust--A major fault in the eastern part of Inyo County, California, *U.S. Geologic Survey Professional Paper* **550-D**, D23-D34.

Sylvester, A.G. and Babcock, J.W. (1975) Regional significance of multiphase folding in the White-Inyo Range, eastern California, *Geological Society of America Abstracts with Programs*, **7**, 1289.

Sylvester, A.G., Miller, C.F. and Nelson, C.A. (1978a) Monzonites of the White-Inyo Range, California and their relations to the calc-alkaline Sierra Nevada batholith, *Geological Society of America Bulletin*, **89**, 1677-1687.

Sylvester, A.G., Oertel, G., Nelson, C.A. and Christie, J.M. (1978b) Papoose Flat pluton: A granitic blister in the Inyo Mountains, eastern California, *Geological Society of America Bulletin*, **89,** 1205-1219.

SUPERIMPOSED COMPRESSIONAL AND EXTENSIONAL TECTONICS IN THE KARKONOSZE-IZERA BLOCK, NE BOHEMIAN MASSIF

S. MAZUR & R. KRYZA
Institute of Geological Sciences, University of Wrocław
50-205 Wrocław, Cybulskiego 30, Poland

Abstract - The metamorphic envelope of the Karkonosze granite in the Karkonosze-Izera block of the western Sudetes is subdivided into three main tectonic units. Structurally upwards these are: (1) Izera-Karkonosze (gneisses and mica schists), (2) South Karkonosze (phyllites and metabasites), and (3) Leszczyniec (metabasites and gneisses) units. All three units experienced contrasting metamorphic paths.

Three ductile deformation events (D_1, D_2, D_3) have been documented in the area. Foliation S_1 and mineral lineation L_1 are the dominant structures in the block except for the mica schists of the Izera-Karkonosze unit. In the mica schists, the D_1 fabric is almost entirely obliterated by the D_2 structures. Deformation D_3 affected only the eastern margin of the block and produced local folding of the earlier structures.

The evolution of the Karkonosze-Izera block involved two main stages. The D_1 compressional event was associated with overthrusting and juxtaposition of the major tectonic units. The direction of tectonic transport was to NW in the two lower (Izera-Karkonosze and South Karkonosze) units and, probably, to NNE in the uppermost (Leszczyniec) unit. The D_1 event was pre-dated by the high pressure/low temperature metamorphism of the South Karkonosze and Leszczyniec units and resulted in their tectonic uplift. The D_2 extensional event involved ESE-directed collapse on the eastern margin of the block related to WNW-ESE extension. The progressive extension promoted the uplift and exhumation of the lowermost Izera-Karkonosze unit and emplacement of the Karkonosze pluton. The ductile shearing in the metamorphic complexes was accompanied by brittle faulting at a higher intersection level to the east. The normal faults controlled the subsidence of the Intra-Sudetic basin.

1. Introduction

The Karkonosze-Izera block, western Sudetes (SW Poland), was considered by Kossmat (1927) as the eastern extension of the German Saxothuringian zone (Fig. 1). The block comprises a large Variscan granite pluton in the core (the Karkonosze granite), mantled by a metamorphic envelope. The metamorphic envelope has been considered variously as:

(I) a stratigraphically uniform Upper Proterozoic and Lower Paleozoic sequence, metamorphosed and deformed during a Late Caledonian orogenic event (Berg, 1912; Maska, 1954; Teisseyre, 1971,1973; Oliver et al., 1993);

(II) an assemblage of a few distinct units of Precambrian and Early Paleozoic age, separated by unconformities (Chaloupsky, 1965, 1989) or tectonic contacts (Kodym and Svoboda, 1948; Oberc, 1960; Szałamacha and Szałamacha, 1958, 1968).

New data from the eastern margin of the Karkonosze-Izera block (Kryza and Mazur, 1995) indicate different metamorphic paths for the main tectonic units. This paper presents a new model of the tectono-metamorphic evolution of the block which is consistent with these results. The juxtaposition of the major units is explained by

O. Oncken and C. Janssen (eds.), Basement Tectonics 11, 51–66.

overthrusts followed by gravitational collapse. The work is based on detailed structural studies within the eastern margin of the block and reconnaissance data from the remaining parts.

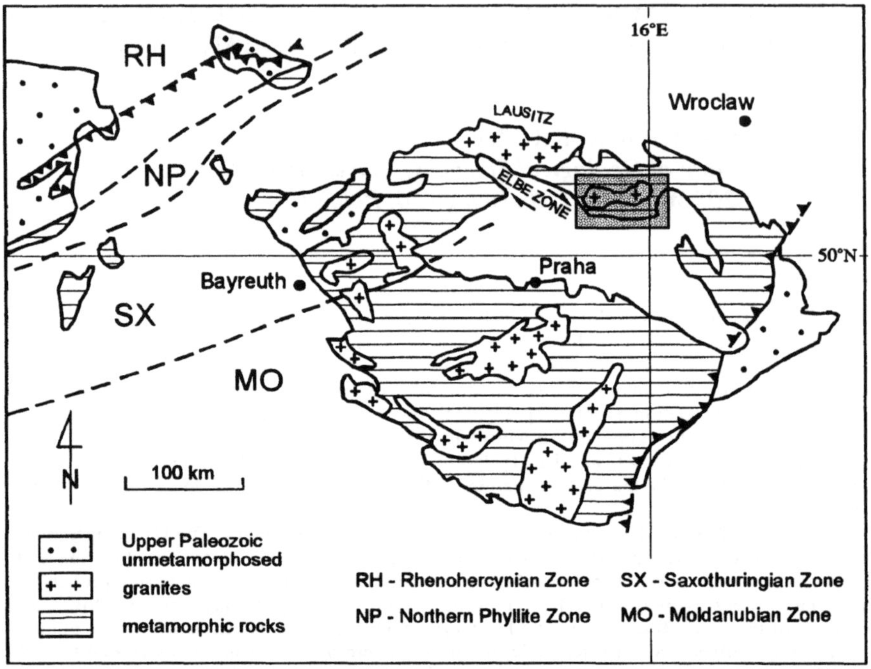

Figure 1. The main structural units of the Variscan belt of central Europe (modified after Franke, 1989). The location of the study area is indicated by the shaded rectangle.

2. Geological setting and previous interpretations

The Karkonosze-Izera block is separated, in the north-east, by the strike-slip Intra-Sudetic fault (Aleksandrowski, 1990; Aleksandrowski et al., 1993; Oliver et al., 1993) from the Paleozoic Kaczawa complex (Fig. 2). This complex has experienced blueschist and subsequent greenschist facies metamorphism (Kryza et al., 1990). To the west, the block is in contact with the Lusatian block in Germany (Fig. 1). Carboniferous through Permian continental sediments, starting with a thick sequence of Visean conglomerates, were deposited east of the Karkonosze-Izera block in the western part of the Intra-Sudetic basin. The south-western boundary of the block is represented by a sedimentary and tectonic (Lusatian fault) contact with Upper Carboniferous-Permian volcano-sedimentary and Cretaceous sedimentary sequences.

The geology of the Karkonosze-Izera block and previous interpretations are summarized by Mierzejewski and Oberc-Dziedzic (1990). Several different hypotheses were proposed to explain the evolution of the block. Closs (1925) emphasized its dome-like structure which was, in his opinion, due to the uplift contemporaneous with the main folding event. He suggested that this event, preceding the emplacement of the Karkonosze granite, took place in the Early Carboniferous. Kodym and Svoboda (1948) suggested that the block is a Caledonian nappe structure and that the higher-grade metamorphic gneisses and mica schists were S-directed overthrust over the lower-grade phyllites and metabasites. The kinematic reconstruction by these authors was based mainly on the predominant southern vergence of folds. According to Maska (1954), the metamorphic envelope of the Karkonosze granite represents a single continuous sequence deformed during the Caledonian orogeny. The foliation and lineation were thought to be reoriented by a N-S striking monocline extending along the eastern margin of the block. Oberc (1961) assumed several phases of folding, during the Late Proterozoic, due to N-S shortening followed by westward over-thrusting of the easternmost Leszczyniec unit (Fig. 2). Chaloupsky (1965, 1989) was of the opinion that each of the major lithostratigraphic subdivisions of the Proterozoic and early Paleozoic age was deposited on a previously metamorphosed and eroded basement. Teisseyre (1973) distinguished four deformation events in the eastern margin of the block which he inferred to be related to the latest phases of the Caledonian orogeny. The internal structure of the Karkonosze pluton has been studied by Closs (1925), Petrascheck (1943), Mierzejewski (1973), Diot and Mierzejewski (1994) and Diot et al. (1994) and several detailed models were proposed.

The metamorphic envelope of the Karkonosze granite can be subdivided into three major tectonic units (Fig. 2): (a) gneisses and mica schists with very minor metavolcanic rocks (here defined as the Izera-Karkonosze unit), (b) phyllites and metabasites (South Karkonosze unit), and (c) metabasites and gneisses (Leszczyniec unit). This tectonic subdivision roughly corresponds to the earlier lithostratigraphic schemes for the southern and eastern part of the Karkonosze-Izera block (Chaloupsky, 1965, 1989; Teisseyre, 1971, 1973). The present subdivision does not include, however, the Middle Devonian-Lower Carboniferous low-grade metamorphic sequences of the Jested area (Fig. 2) which are not considered in this paper. These sequences form either a separate tectono-stratigraphic unit (Chaloupsky, 1989) or constitute the upper part of the South Karkonosze unit (Chlupac and Hladil, 1992).

The data from the eastern part of the Karkonosze-Izera block indicate that the gneisses and mica schists experienced medium pressure (MP) metamorphism, generally below the amphibolite facies, followed by a relatively high temperature (HP) and low pressure (LP) event (Kryza and Mazur, 1995). The metabasites of the South Karkonosze unit bear evidence of HP/LT blueschists facies metamorphism overprinted by a MP greenschist facies episode (Wieser, 1978; Chab and Vrana, 1979; Guiraud and Burg, 1984; Patocka et al., 1994; Kryza and Mazur, 1995). The metabasites and gneisses of the Leszczyniec unit underwent relatively HP metamorphism up to epidote-amphibolite facies conditions (Kryza and Mazur, 1995).

The protolith ages and the age of metamorphism in the Karkonosze-Izera block are poorly constrained. Recent U-Pb dating has yielded ages of ca 500 Ma for the gneisses of the Izera-Karkonosze unit (Oliver et al., 1993; Korytowski et al., 1993; Kröner et al., 1994) and for the metabasites of the Leszczyniec unit (Oliver et al., 1993). Rb-Sr whole rock analyses of the Izera gneisses yielded a date of ±460 Ma (Borkowska et al., 1980). The only biostratigraphically documented strata are phyllites of the South Karkonosze unit which contain Silurian graptolites (Perner, 1919; Chlupac, 1953). And, finally, the Karkonosze granite has been dated, using Rb-Sr whole rock method, at ±329 Ma (Pin et al., 1987; Duthou et al., 1991). The granite shows no clear evidence of ductile deformation and, locally, displays cross-cutting relationships against its cover.

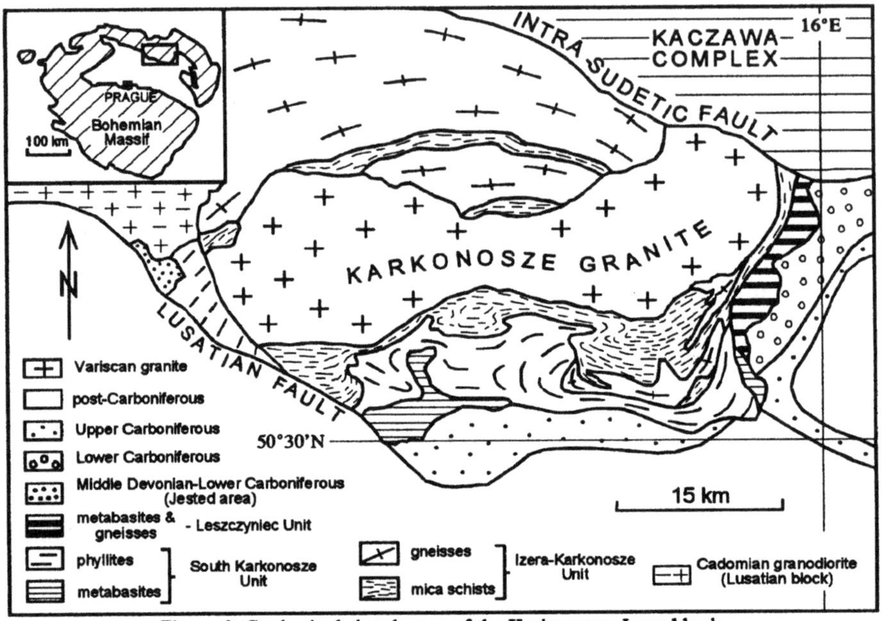

Figure 2. Geological sketch map of the Karkonosze-Izera block.

3. Deformation

Three ductile deformation events (D_1, D_2, D_3) have been documented in the eastern part of the Karkonosze-Izera block (Mazur, 1993; Mazur and Kryza, 1994). They roughly correspond to the three main deformations distinguished by Teisseyre (1973). Foliation S_1 and mineral lineation L_1 are the dominant structures in the study area except for the mica schists of the Izera-Karkonosze unit. In the mica schists, the D_1 fabric is almost entirely obliterated by the D_2 structures but relics of the earlier S_1 foliation are still preserved as inclusion trails in plagioclase and garnet porphyroblasts. Deformation D_3 is evident mainly in the eastern margin of the block and the adjacent Lower Carboniferous conglomerates. In the metamorphic rocks, deformation D_3 resulted in local folding of the earlier structures.

3.1. D_1 STRUCTURES

3.1.1. Izera-Karkonosze unit

The gneisses of the Izera-Karkonosze unit are coarse-grained rocks, typically with augen texture developed around variably deformed and recrystallized K-feldspar megacrysts. S_1 foliation is defined by the parallel orientation of mica plates, flattened feldspar megacrysts and alternating quartzo-feldspatic and mica layers. The quartzofeldspar megacrysts and alternating quartzo-feldspatic and mica layers. The quartzo-feldspatic aggregates are often differentiated into quartz and feldspar laminae. The predominant oblate shape of the K-feldspar megacrysts in the gneisses indicates a flattening type of deformation. The increasing strain is recorded by progressive flattening of these megacrysts. Higher strained gneisses are characterized by more distinct planar textures and lower K-feldspar contents. The textures of very coarse-grained massive varieties of the gneisses resemble those of a granite.

The S_1 foliation in the mica schists is, in many places, preserved as inclusion trails in plagioclase and garnet porphyroblasts. In the XZ section, the trails are usually rectilinear or only slightly curved while in the YZ section they are distinctly bent or, in places, define microfolds.

The S_1 foliation in the gneisses has a stretching lineation L_1 which is manifested as elongated feldspar grains, quartz ribbons and mica aggregates. The higher strained gneisses are well-lineated and display distinct rotational criteria suggesting deformation by simple shear. Large masses of less strained gneisses display a nearly symmetrical fabric which suggests an approximately coaxial deformation.

The composition of the garnet cores and plagioclase porphyroblasts in the gneisses and mica schists of the Izera-Karkonosze unit indicates that the unit underwent upper greenschists facies metamorphism during D_1 deformation (Kryza & Mazur, 1995). The cores of the plagioclase grains are usually composed of albite while those of the garnets are relatively spessartine-rich and almandine-poor.

3.1.2. South Karkonosze unit

The metabasites of the South Karkonosze unit comprise a wide range of rocks, from greenstones to metadiabases and metagabbros. The most widespread variety is fine-grained to aphanitic, foliated greenstone, rich in fibrous, pale-green actinolite, albite, epidote and chlorite. Sodic amphibole, and Mn-rich garnet are occasionally present. The sodic amphibole is blue or violet to colourless glaucophane-crossite (Wieser, 1978; Chab and Vrana, 1979; Guiraud and Burg, 1984; Patocka et al., 1994; Kryza and Mazur, 1995). The S_1 foliation is bounded by the parallel orientation of amphibole prisms, elongated albite porphyroblasts and chlorite plates. Thin lamination, generally S_1-parallel, is common in some metabasites. The lamination is defined by the alternation of light albite (± quartz or calcite) laminae, and dark ones composed mainly of amphibole, epidote and chlorite. Locally, the lamination is deformed by tight microscopic F_1 folds. In fold hinges, foliation S_1 is oblique to the lamination.

The L_1 lineation in the metabasites is defined by the parallel alignment of the amphibole prisms while in the glaucophane-bearing metabasites sodic amphibole crystals define lineation L_1. Cores of crystals composed of glaucophane or crossite are often rimmed by pale-green actinolite. The alignment of actinolite defines L_1 in the greenschists.

The phyllites are fine-grained, well-foliated rocks. The S_1 foliation is defined in these rocks by the alternation of thin, continuous, quartz and white mica layers and parallel orientation of white mica plates and elongated albite porphyroblasts. The L_1 lineation in the phyllites is manifested by the parallel alignment of white mica and small crenulations.

3.1.3. Leszczyniec unit

The textures of the metabasites in the Leszczyniec unit vary, the most widespread variety being fine-grained schists. The S_1 foliation in those rocks is defined by the parallel orientation of amphibole prisms and elongated albite porphyroblasts. Locally, relict layering is preserved as thin, leucocratic layers enriched in albite or quartz. In places, the layering is oblique to the foliation. Medium-grained massive varieties of the metabasites display a well-preserved gabbroic texture. Euhedral crystals of amphibole often exhibit distinct zonation: pale-green actinolite in the core and bluish-green hornblende in the rim. The L_1 lineation in the metabasites is defined by the parallel alignment of amphibole prisms: bluish-green hornblende and, in places, actinolite. In some samples pale-green actinolite in the core of a crystal is rimmed by synkinematic hornblende.

The gneisses of the Leszczyniec unit (so called Paczyn gneisses) form several sill-like bodies, a few tens to few hundreds of meters thick, interlayered with the metabasites. The gneisses include a wide range of rock types varying in texture and composition. The most striking feature is the lack of K-feldspar. The felsic varieties display a coarse-grained augen texture developed around variably recrystallized quartz megacrysts. The S_1 foliation is manifested as parallel megacrysts and white mica plates and, locally, by alternating white mica and quartz-albite layers. The increasing strain is recorded in gneisses by progressive flattening and recrystallization of quartz megacrysts. Higher strain values resulted in transition of the augen gneisses into thinly-laminated varieties. In contrast, low strained gneisses grade into massive granitic rocks. In the felsic varieties of gneisses, lineation L_1 is defined by the parallel alignment of white mica and elongated quartz megacrysts or aggregates. The generally asymmetric fabric reflects simple shear during D_1 deformation. It is only quartz megacrysts in the felsic gneisses that bear record of coaxial flattening. This may have resulted from strain partitioning (Lister & Williams, 1983).

The hornblende-rich gneisses are medium- to fine-grained, and massive to weakly foliated rocks. In places, the euhedral hornblende and plagioclase grains define a dioritic texture. The S_1 planes contain parallel hornblende crystals and elongated albite blasts or aggregates. The amphibole cores made of hornblende usually display greenish-brown pleochroism and are rimmed by pale bluish-green actinolite. The parallel alignment of amphibole defines lineation L_1 in these gneisses.

3.1.4. Orientation of D_1 structures

The foliation in the Karkonosze-Izera block roughly parallels the lithologic contacts and defines a large antiform (Fig. 3). In the southern and eastern part of the block, the foliation displays an arcuate pattern. Its dip, which is generally steep in this area, gradually decreases toward the southern boundary of the granite. North of the pluton, the foliation planes are steep and show a curvate pattern (Fig. 3).

The stretching lineation L_1 trends WNW-ESE almost everywhere in the block except for the Leszczyniec unit where it trends NNE-SSW (Fig. 4). In the easternmost part of the South Karkonosze unit, at the eastern margin of the block (Fig. 2), L_1 changes orientation from E-W to NE-SW and NW-SE.

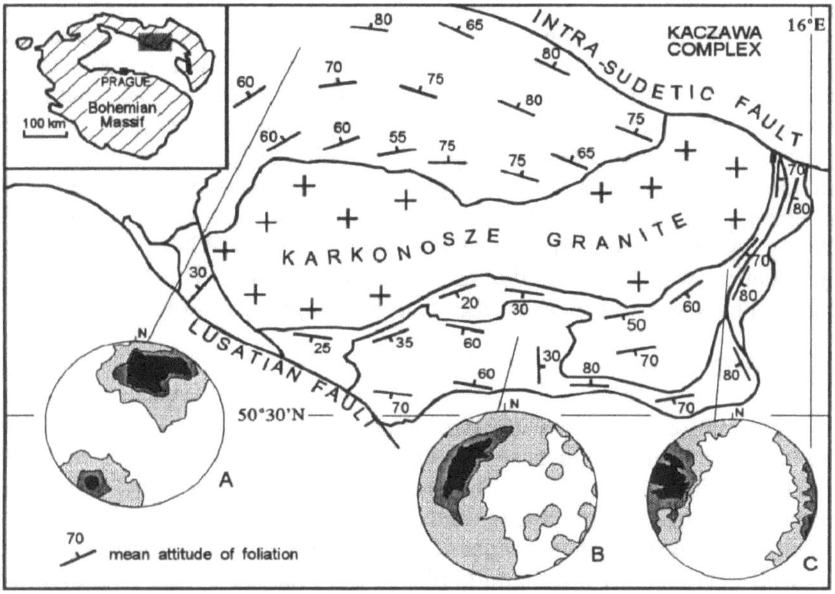

Figure 3. Orientation of foliation in the Karkonosze-Izera block.
Foliation plotted on Schmidt stereograms (lower hemisphere) for: (A) northern flank of the block (1450 measurements) after Oberc (1967), (B) southern flank of the block (1400 measurements) after Maska (1954), and (C) eastern margin of the block (1050 measurements) after Mazur (1993). Contours (per 1% area): 0.5, 2, 3, max. 6% (A); 0.25, 1, 3, max. 5% (B); 1, 2, 3, max. 5% (C).

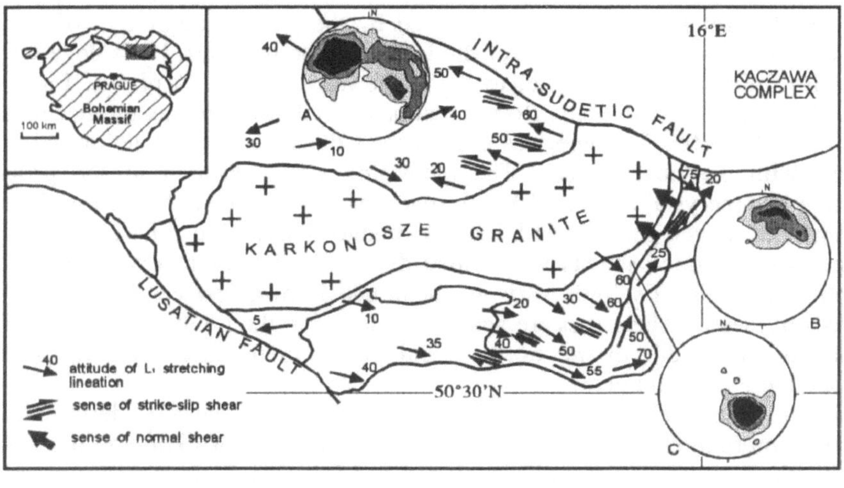

Figure 4. Orientation of the L_1 stretching lineation and associated sense of shear.
Lineation plotted on Schmidt stereograms (lower hemisphere) for: (A) northern flank of the block (650 measurements) after Oberc (1967), (B) Leszczyniec unit (240 measurements) after Mazur (1993), and (C) eastern margin of the Izera-Karkonosze unit (250 measurements) after Mazur (1993). Contours (per 1% area): 1, 3, 7, max. 14%.

The asymmetry of the D_1 fabric provides evidence of a widespread simple shear component in the rocks of the study area. The most common kinematic indicators related to D_1 are: (a) asymmetric recrystallized tails around porphyroblasts (Simpson and Schmid, 1983), (b) asymmetric pressure shadows (Simpson and Schmid, 1983; Etchecopar and Malavieille, 1987), (c) rotated porphyroblasts (Passchier and Simpson, 1986), (d) extensional shear bands (Gapais and White, 1982). These indicators show a sinistral shear north of the Karkonosze pluton and a dextral shear to the south of the pluton. The L_1 lineation is subhorizontal and foliation dips to the north and south, respectively. At the eastern margin of the block, a top to WNW shear on the ESE-dipping foliation was found in the Izera-Karkonosze unit. The Leszczyniec unit is characterized by a sinistral shear on the steep ESE-dipping foliation (Fig. 4).

3.2. D_2 STRUCTURES

The D_2 deformation structures are particularly well developed in the rocks of the Izera-Karkonosze unit. In mica schists the deformation produced a new dominant fabric. The distinct S_2 foliation of these rocks is defined by the small-scale differentiation of quartz and micas, and by the parallel orientation of mica plates. Intergrowths of biotite with subsequent chlorite are very common. In places, S-C fabric is well-developed. Both S and C planes are bounded by the alignment of micas and by mineral segregation into thin quartz and mica layers. Synkinematic porphyroblasts of plagioclase and, locally, garnet are common components of the mica schists. The porphyroblasts usually contain relics of S_1 preserved as inclusion trails. The distinct zonation of plagioclase and garnet porphyroblasts indicates gradual decrease of P/T gradient during their growth (Kryza & Mazur, 1995). This is consistent with the zonation of white mica plates parallel to S_2. The asymmetric pressure shadows around garnet are composed of chlorite, white mica and quartz.

The L_2 stretching lineation is outlined by the parallel alignment of micas and small crenulations. In the gneisses of the Izera-Karkonosze unit, L_2 is developed on a composite foliation S_{1+2} parallel to S_2 in the mica schists. In places, L_2 is superimposed on the still visible L_1. Both lineations are oblique to one another making an angle of 20° to 40°. The gneisses experienced the most intense deformation along contacts with the mica schists. The increasing strain resulted in the progressive flattening and recrystallization of K-feldspar megacrysts. The deformation thinned the layering of the augen gneisses and, locally, produced mylonitic zones. The foliation in the gneisses and mica schists was folded during the D_2 event resulting in folds of variable scales. The F_2 folds vary in magnitude from a few centimetres to several hundreds of meters. Their axes are generally parallel to the L_2 stretching lineation and their axial planes dip to the north. Axial planar crenulation cleavage is usually developed in F_2 fold hinges.

In the South Karkonosze and Leszczyniec units, the D_2 deformation was apparently more intense along their contacts with the Izera-Karkonosze unit. Near the contacts, the L_2 lineation is well-developed and foliation S_2 is locally present in the phyllites. F_2 folds are also very common in these zones. L_2 is manifested there as the parallel alignment of white mica, small crenulations and the intersection of foliation with crenulation cleavage in the hinges of F_2 folds. With increasing distance away from the boundaries of the Izera-Karkonosze unit, the effects of the D_2 deformation decrease. East of the Karkonosze granite, the ductile shearing in the Izera-Karkonosze unit is replaced by brittle-ductile deformation in the Leszczyniec unit, and by brittle normal faults at the boundary with the Intra-Sudetic basin.

Figure 5. Orientation of the L_2 stretching lineation and associated sense of shear. Lineation plotted on Schmidt stereograms (lower hemisphere) for the eastern margin of the block (300 measurements) after Mazur (1993). Contours (per 1% area): 1, 3, 7, max. 14%. Attitude of magnetic lineation in the granite after Diot and Mierzejewski (1994).

The L_2 lineation trends WNW-ESE throughout the Karkonosze-Izera block (Fig. 5). Its plunge, which is steep within the eastern margin of the block, gradually decreases westward. Kinematic indicators such as: (a) S-C relationships (Berthe et al., 1979; Lister and Snoke, 1984), (b) asymmetric recrystallized tails around feldspar porphyroblasts (Simpson and Schmid, 1983), (c) asymmetric pressure shadows (Simpson and Schmid, 1983; Etchecopar and Malavieille, 1987), and (d) extensional shear bands (Gapais and White, 1982) show normal sense of shear on the steep ESE-dipping foliation within the eastern margin of the block. Evidence for sinistral wrench/normal shear on the S-dipping foliation has been observed along the southern flank of the block (locally overprinted by a later phase of dextral shear) (Fig. 5). In the northern flank, an approximately strike-slip dextral shear on a steep N-dipping foliation is suggested by kinematic indicators, in particular along the Intra-Sudetic fault (Aleksandrowski et al., 1993). However, a subsequent brittle-ductile sinistral shear is also evidenced in that area by a kilometre-scale strike-slip duplex (Aleksandrowski et al., 1993) and by passive bending of the foliation in the NE corner of the block.

A local WNW-ESE trending, non-penetrative lineation, defined mainly by the alignment of the feldspar megacrysts (Cloos, 1925; Mierzejewski, 1973), parallels L_2 in the metamorphic envelope of the Karkonosze granite (Cloos, 1925). The existence of a weak deformational fabric in the granite, ascribed to D_2, has recently been confirmed by studies of magnetic fabric (Diot et al., 1994).

3.3. D_3 STRUCTURES

Within the eastern margin of the Karkonosze-Izera block, the foliation becomes increasingly steeper toward the east. Its dip changes from 60°E close to the granite, through vertical, to 70°-80°W at the boundary with the Intra-Sudetic basin. The reorientation of the foliation defines a longitudinal monocline with smaller scale folds that have subhorizontal N-S trending axes. The monocline is also developed in the adjacent Lower Carboniferous conglomerates (Dziedzic and Teisseyre, 1990). Farther to the east, the folded Visean strata are uncomformably overlain by Upper Carboniferous sediments (Dziedzic and Teisseyre, 1990). Subhorizontal slickensides and tectonic breccias along the eastern margin of the block may be the result of dextral brittle-ductile strike-slip displacement during D_3 deformation.

4. Discussion

The inferred structural sequence of the tectonically stacked units is based mainly on field relationships from the eastern margin of the Karkonosze-Izera block. This part of the block is represented by a regional down-to-the-east, dip-slip shear zone which developed during the D_2 event (Fig. 6). Since the tectonic units exposed in the footwall (upthrown side) of the shear zone are expected to be structurally lower, we assume the following upward sequence (Fig. 7): the Izera-Karkonosze unit, the South Karkonosze unit and the Leszczyniec unit. Our interpretation is different from that of Kodym and Svoboda (1948), which has recently been accepted by the "Giessen Group" (W. Franke, pers. comm. 1995). They infer that the outcrop pattern in the southern part of the Karkonosze-Izera block indicates that the South Karkonosze unit is overlain by the Izera-Karkonosze unit. However, the Izera-Karkonosze unit in the same area has also been interpreted to form the core of a recumbent south-vergent fold (Oberc, 1961) and to represent the basement of the South Karkonosze unit (Chaloupsky, 1965, 1989). In the light of conflicting evidence and interpretations, the tectonic sequence of the main units remains controversial but our model seems to be favoured by the geometric and structural data.

Before the onset of D_1 deformation, the rocks of the three major tectonic units were under different P-T conditions (Fig. 8). The early HP/LT event in the South Karkonosze and Leszczyniec units, which is interpreted as the pre-collisional metamorphic phase, must have been connected with their deep burial, possibly due to Early Paleozoic subduction. In contrast, the Izera-Karkonosze unit was metamorphosed under medium pressure and low- to moderate temperatures.

During D_1 metamorphic paths of the three units converged indicating their tectonic juxtaposition. The changing parageneses in the South Karkonosze and Leszczyniec units suggest decreasing pressure during D_1. Therefore, the D_1 deformation appears to have been related to the uplift history of both units. The Izera-

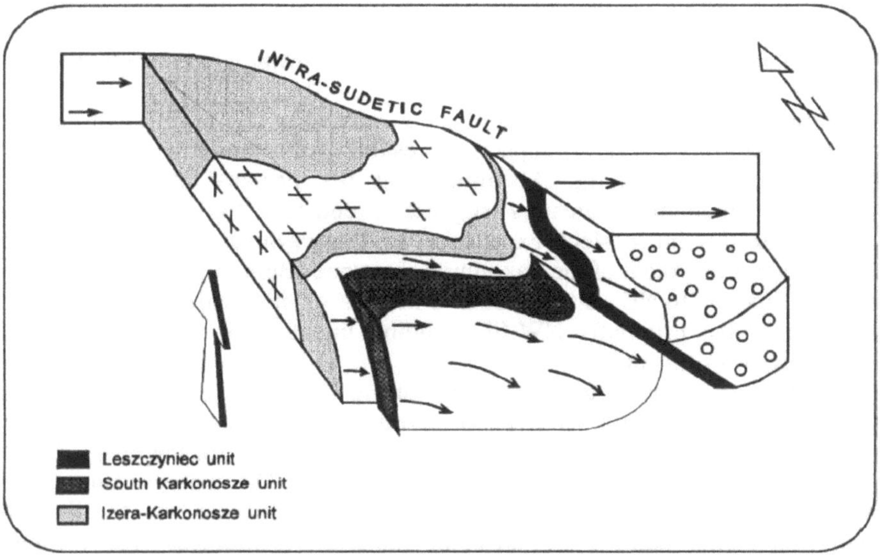

***Figure* 6. Schematic block diagram showing the kinematics during the late phase of the D_2 extensional deformation in the eastern part of the Karkonosze-Izera block. Arrows indicate displacement vectors.**

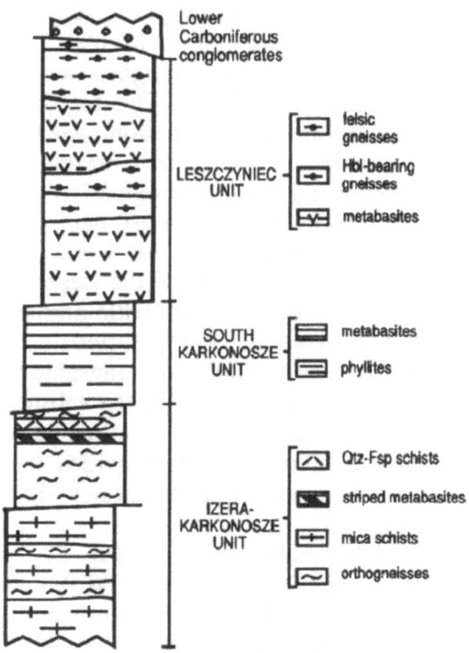

***Figure* 7. Structural sequence in the SE part of the metamorphic envelope of the Karkonosze granite (after Kryza and Mazur, 1995 - simplified).**

Karkonosze unit most probably remained under MP regime during D_1. The location of the HP South Karkonosze and Leszczyniec units above the MP Izera-Karkonosze unit, and of the higher grade Leszczyniec unit above the lower grade South Karkonosze unit, may be the result of overthrusting.

The mutually inconsistent senses of shear, related to D_1, in different parts of the Karkonosze-Izera block (Fig. 4) apparently resulted from reorientation of the S_1 due to the superimposed doming. Assuming that the primary attitude of S_1 was subhorizontal, kinematic indicators suggest a constant top to WNW sense of displacement in the entire block except for the Leszczyniec unit. The NW-directed tectonic transport is typical of the Saxothuringian zone (Behr et al., 1984). However, it is not clear if the top to NE sense of shear in the Leszczyniec unit resulted from a later reorientation of the originally NW-SE trending lineation or from the NE-directed tectonic transport of this unit.

The direction of the tectonic transport during D_2 was nearly opposite to that of the D_1 event. The reversed sense of shear during D_2 was inferred from standard kinematic criteria; however, the opposite sense of shear was not the only criteria that we used to distinguished between D_2 and D_1 structures. The superposition of D_2 and D_1 events in the study area is documented by the occurrence of two, locally oblique, stretching lineations defined by different mineral assemblages. Also, kinematic indicators produced by D_1 and D_2 are associated with the effects of different mineral parageneses. This evidence suggests that the mutually opposite senses of shear observed in the rocks of the Karkonosze-Izera block resulted from two separate deformation events that took place under different metamorphic conditions. Nevertheless, in some locations, the recorded opposite sense of shear may reflect data collected on the opposite limbs of lineation-parallel folds.

The kinematics of the D_2 deformation suggests an ESE-directed extensional collapse in the Karkonosze-Izera block (Fig. 6). This top to the ESE shearing, which may have started on the originally flat laying foliation, was followed, or accompanied, by doming. Normal displacements on the eastern margin of the Karkonosze-Izera block resulted in westward-increasing uplift of progressively lower tectonic units. The Izera-Karkonosze unit, exhumed in the innermost part of the hanging wall, was overlain to the east by the South Karkonosze and Leszczyniec units (Fig. 7). The Visean conglomerates cropping out to the east of the Leszczyniec unit correspond to the hanging wall of this fault. The kinematics of brittle normal faults which bound the conglomerates on the west (Teisseyre, 1975; Dziedzic & Teisseyre, 1990) is equivalent to the ductile normal shearing in the metamorphic rocks during D_2. Normal displacements on these faults must have controlled the subsidence of the basin, where a 4000 m thick coarse-clastic sequence of Visean continental sediments was deposited. The occurrence of ESE-dipping normal faults provides the evidence for regional WNW-ESE extension during D_2, and it demonstrates that normal displacements continued at shallower depths.

The contrasting metamorphic paths during D_2 (Fig. 8) confirm the tectonic uplift of the inner Izera-Karkonosze unit. The relatively HT/LP metamorphism experienced by this unit was probably the effect of both rapid decompression of the uplifted rocks and the thermal influence of the Karkonosze intrusion.

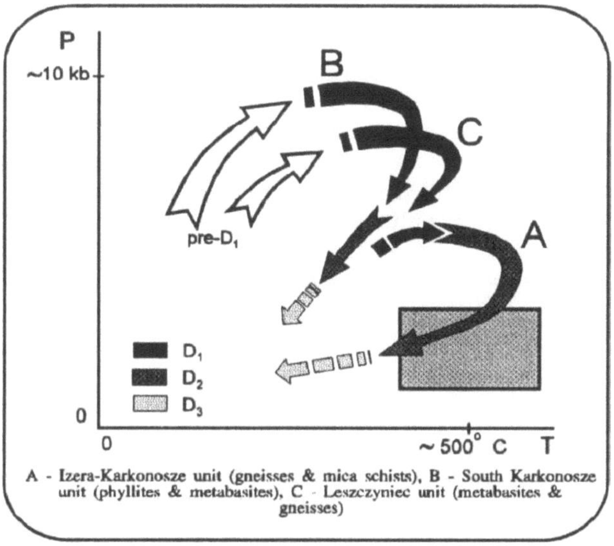

A - Izera-Karkonosze unit (gneisses & mica schists), B - South Karkonosze unit (phyllites & metabasites), C - Leszczyniec unit (metabasites & gneisses)

***Figure* 8. Tentative P-T paths for particular tectonic units of the Karkonosze-Izera block (shaded box represents the approximate field of LP-HT metamorphism connected with the Karkonosze granite intrusion).**

The emplacement of the Karkonosze granite must have taken place during the D_2 extensional deformation. The weak fabric in the granite indicates that the emplacement of the pluton occurred during the last phases of extension. The lack of a thermal aureole in the rocks in contact with the granite across the Intra-Sudetic fault, at the NE corner of the pluton, indicates that the movements along the fault continued after the intrusion.

The development of the monocline along the eastern margin of the Karkonosze-Izera block during the D_3 event was associated with the continued stretching in the WNW-ESE direction. D_3 probably marked the end of the first phase of subsidence of the Intra-Sudetic basin.

The age of the D_1 event is poorly constrained. The lower limit for this deformation is established by the paleontologically documented Silurian strata in the South Karkonosze unit (Perner, 1919; Chlupac, 1953). The timing of D_2 event is roughly indicated by the Visean age of the molasse sedimentation in the Intra-Sudetic basin. The Rb/Sr age of the Karkonosze granite estimated at 329 ±17 Ma (Pin et al., 1987; Duthou et al., 1991) confirms this interpretation. The D_3 deformation may partly postdate the Visean sedimentation.

5. Conclusions

The evolution of the Karkonosze-Izera Block involved two main stages:

1. D_1 compressional event associated with overthrusting. This deformation was pre-dated by the HP/LT metamorphism of the two upper units which were subsequently uplifted during D_1 and emplaced above the MP unit. The direction of

tectonic transport was to NW in the two lower units and probably to the NE in the uppermost unit.

2. D_2 extensional event comprised ESE-directed collapse on the eastern margin of the block related to WNW-ESE extension. The progressive extension promoted the uplift and exhumation of the lowermost MP unit on the west. The ductile shearing in the metamorphic complexes was accompanied by brittle faulting higher up in the section to the east which controlled subsidence of the Intra-Sudetic basin. During the last phases of D_2 the Karkonosze pluton was emplaced.

***Acknowledgements** - The study was financed from the Polish Research Committee KBN grants nos. 6 0434 91 01 and 6 6194 92 03.*

6. References

Aleksandrowski, P. (1990) Early Carboniferous strike-slip displacements at the northeast periphery of the Variscan belt in central Europe, *Int. Conf. on Paleozoic Orogens in Central Europe (IGCP Program 233: Terranes in the Circum-Atlantic Paleozoic Orogens)*, Göttingen-Giessen, Abstracts, 7-10.

Aleksandrowski, P., Żaba, J., and Ciukszo, B. (1993) Changing sense of strike-slip motion and mode of failure on a Variscan terrane boundary: Intra-sudetic fault zone, SW Poland, *Terra Abstracts* **5** (2), 1.

Behr, H.-J., Engel, W., Franke, W., Giese, P., and Weber, K. (1984) The Variscan belt in central Europe: main structures, geodynamic implications, open questions, *Tectonophysics* **109**, 15-40.

Berg, G. (1912) Die krystallinen Schiefer des östlichen Riesengebirges, *Abh. Preuss. Geol. L.-A.* **28**, 188, Berlin.

Berthé, D., Choukroune, P., and Jegouzo, P. (1979) Orthogneiss, mylonite and non-coaxial deformation of granites: the example of the South Armorican shear zone, *J. Struct. Geol.* **1**, 31-42.

Borkowska, M., Hameurt, J., and Vidal, P. (1980) Origin and age of Izera gneisses and Rumburk granites in the Western Sudetes, *Acta Geol. Polon.* **30**, 121-145.

Chab, J. and Vrana, S. (1979) Crossite-actinolite amphiboles of the Krkonose-Jizera crystalline complex and their geological significante, *Vestnik Ustr. Ust. geol.* **54** (3), 143-150.

Chaloupsky, J. (1965) Metamorphic development of the Krkonose crystalline complex, *Krystalinikum* **3**, 31-54.

Chaloupsky, J. (ed.) (1989) *Geologie Krkonos a Jizerskich hor (Geology of the Karkonose and Jizerske hory Mts.)*, Ustredni Ustav Geologicky, 288 pp.; Praha.

Chlupac, I. (1953) Finding of graptholites in metamorphosed Silurian of Żelezny Brod in Podkrkonosi, *Vest. Ustr. Ust. Geol.* **28**, 213-215.

Chlupac, I., and Hladil, J. (1992) New Devonian occurrences in the Jested Mountains, North Bohemia, *Casopis pro mineralogii a geologii* **37** (3), 185-191.

Cloos, H. (1925) *Einführung in die tektonische Behandlung magmatischer Erscheinungen (Granittektonik)*, I Spez. Teil. Das Riesengebirge in Schlesien. - 194 pp.; Berlin.

Diot, H., Mazur, S., and Couturie, J. P. (1994) Magmatic structures in the Karkonosze granite and their relation to tectonic structures in the eastern metamorphic cover, *Int. Conf., Igneous activity and metamorphic evolution of the Sudetes area*, Wrocław, Abstracts, 36-39.

Diot, H. and Mierzejewski, M., P. (1994) The magnetic fabric in the Karkonosze granite, *Int. Conf., Igneous activity and metamorphic evolution of the Sudetes area*, Wrocław, Abstracts, 34-35.

Duthou, J. L., Couturie, J. P., Mierzejewski, M. P., and Pin, C. (1991) Rb/Sr age of the Karkonosze granite on the base of the whole rock method, *Geol., Review* **2**, 75-79.

Dziedzic, K., and Teisseyre, A. K. (1990) The Hercynian molasse and younger deposits in the Intra-Sudetic Depression, SW Poland, *N. Jb. Geol. Paläont. Abh.* **179** (2/3), 285-305.

Etchecopar, A., and Malavieille, J. (1987) Computer models of pressure shadows: a method for strain measurement and shear-sense determination, *J. Struct. Geol.* **9**, 667-677.

Franke, W. (1989) Tectonostratigraphic units in the Variscan belt of central Europe, *Geol. Soc. Am. Spec. Paper* **230**, 67-89.

Gapais, D., and White, S. H. (1982) Ductile shear bands in a naturally deformed quartzite, *Textures and Microstructures* **5**, 1-17.

Guiraud, M. and Burg, J. P. (1984) Mineralogical and petrological study of a blueschist metatuff from the Żelezny Brod crystalline complex, Czechoslovakia, *N. Jb. Miner. Abh.* **149** (1), 1-12.

Kodym, O. and Svoboda, J. (1948) Caledonian nappe structure of the Karkonosze and Jizera Mountains, *Sb. St. Geol. Ust.* **15**, 109-160.

Korytowski, A., Dörr, W., and Żelaźniewicz, A. (1993) U-Pb dating of (meta)granitoids in the NW Sudetes (Poland) and their bearing on tectono-stratigraphic correlation, *Terra abstracts* **5** (1), 331.

Kossmat, F. (1927) Gliederung des variszischen Gebirgsbaues. *Abh. sächs. geol. Landesamts* **1**,1-39.

Kröner, A., Jaeckel, P., and Opletal, M. (1994) Pb-Pb and U-Pb ages for orthogneisses from eastern Bohemia: Further evidence for a major Cambro-Ordovican magmatic event, *Journal of the Czech Geol. Society* **39** (1), 61.

Kryza, R., Muszyński, A. and Vielzeuf, D. (1990) Glaucophane-bearing assemblage overprinted by greenschist-facies metamorphism in the Variscan Kaczawa complex, Sudetes, Poland, *J. Metamorphic Geol.* **8**, 345-355.

Kryza, R. and Mazur, S. (1995) Contrasting metamorphic paths in the SE part of the Karkonosze-Izera block (Western Sudetes, SW Poland), *N. Jb. Miner. Abh.* **169** (2), 157-192.

Lister, G. S., and Williams, P. F.(1983) The partitioning of deformation in flowing rock masses, *Tectonophysics* **92**, 1-33.

Lister, G. S., and Snoke, A. W. (1984) S-C mylonites, *J. Struct. Geol.* **6**, 617-638.

Maska, M. (1954) *On tectonic analysis of the crystalline rocks*, Ustredni Ustav Geologicky, 258 pp.; Praha.

Mazur, S. (1993) Structural and metamorphic evolution of the eastern cover of the Karkonosze granite, Unpub. PhD. Thesis, Wrocław University.

Mazur, S. and Kryza, R. (1994) Thrusts and collapse in the Karkonosze-Izera block, NE Bohemian massif, *11th Int. Conf. on "Basement Tectonics '94"*, Potsdam, Abstracts, 98-99.

Mierzejewski, M. P. (1973) Orientation of feldspar megacrystals in relation to the evolution and structures of granitic Karkonosze pluton. *Revue des problemes geologiques des zones profondes de l'encore terrestre en Basse Silesie (XV Session de l'AZOPRO)*, 138-145, Warszawa.

Mierzejewski, M. P., and Oberc-Dziedzic, T. (1990) The Izera-Karkonosze block and its tectonic development (Sudetes, Poland), *N. Jb. Geol. Paläont. Abh.* **179** (2/3), 197-222.

Oberc, J. (1960) Tectonics of the Eastern Karkonosze and their significance in the structure of the Sudetes, *Acta Geol. Polon.* **10** (1), 1-48.

Oberc, J. (1961) An outline of the geology of the Karkonosze-Izera block, Wroclaw Univ. Scientific Papers, IV, ser. B, 8, 139-170.

Oberc, J. (1967) Dispersion of B-lineation in the Izera Mts. crystalline massif (Sudetes Mts.), *Ann. Soc. Geol. Pol.* **37** (3),373-386.

Oliver, G. J. H., Corfu, F., and Krogh, T. E. (1993) U-Pb ages from SW Poland: evidence for a Caledonian suture zone between Baltica and Gondwana, *Journal of Geological Society, London* **150**, 355-369.

Passchier, C. W., and Simpson, C. (1986) Porphyroclast systems as kinematic indicators, *J. Struct. Geol.* **8**, 831-844.

Patocka, F., Pivec, E. and Oliveriova, D. (1994) The mafic blueschists from the Rychory Mountains crystalline complex (Western Sudetes, Bohemian massif): metamorphic development and possible protolith composition, *Int. Conf., Igneous activity and metamorphic evolution of the Sudetes area*, Wrocław, Abstracts, 36-39.

Perner, J. (1919) Silurian in Krkonose, *Cas. Mus. Kral. ces.* **43**, 33-34.

Petraschek, W. E. (1943) Einige Beziehungen zwischen Intrusionstektonik und Lagerstättenverteilung. *Geol. Rdsch.* **34** (1), 38-54.

Pin, C, Mierzejewski M. P., and Duthou, J. L. (1987) Rb/Sr age of the Karkonosze granite from the Huta quarry, *Geol. Review* **10**, 512-517.

Simpson, C., and Schmid, S. M. (1983) An evaluation of criteria to deduce the sense of movement in sheared rocks, *Geol. Soc. Am. Bull.* **94**, 1281-1288.

Szałamacha, M. and Szałamacha, J. (1958) Some remarks on the metamorphism of the crystalline series in the vicinity of Kowary and Niedamirów, *Geol. Review* **8-9**, 353-356.

Szałamacha, M. and Szałamacha, J. (1968) The metamorphic series of the Karkonosze-Izera Mountains block, *Inst. Geol. Bull.* **222**, 33-76.

Teisseyre, J. H. (1971) On the age and sequence of the metamorphic rocks of the Rudawy Janowickie Mountains and Lasocki Range, *Geol. Sudetica* **5**, 165-210.

Teisseyre, J. H. (1973) Metamorphic rocks of the Rudawy Janowickie and Lasocki Range, *Geol. Sudetica* **8**, 7-120.

Teisseyre, A. K. (1975) Sedimentology and paleogeography of the culm alluvial fans in the western part of the Intra-Sudetic basin, *Geol. Sudetica* 9 (2), 1-135.
Wieser, T. (1978) Glaucophane schists and associated rocks of Kopina Mountain. (Lasocki range, Sudetes), *Mineralogia Polonica* 9, 35-56.

LOWER CONTINENTAL CRUST EXHUMED: GRANULITE GRADE TERRANES AND AMPHIBOLITE GRADE SHEAR ZONES IN THE ANABAR SHIELD, POLAR SIBERIA

O.M. ROSEN
Institute of the Lithosphere, R.A.S.
Staromonetny per., 22, Moscow, 109180, Russia

1. Introduction

The lower continental crust (10-40 km deep), modeled from geophysical data and kimberlite xenoliths studies, is assumed to be composed of mafic granulites with seismic velocities as high as P_v=6.35-7.05 km/s, densities of 2.85-3.08 g/cm^3 (Meissner, 1986) and radioactive heat productions of A=0.48-0.53 μW/m^3 (Rudnick, 1992, Kremenetsky et al., 1986). In contrast the upper crust is enriched with highly radioactive granite batholiths and granite-migmatite domes (Taylor and McLennan, 1985) and shows P_v=6.06 km/s, an average density of 2.75 g/cm^3 and A=1.33-1.75 μW/m^3. The lower crust, brought to the surface by upthrusting along decollements, appears to be composed of thinly banded isoclinally folded granulites with intermediate to mafic and ultramafic compositions (Percival et al., 1992, Rutter and Brodie, 1992 and Zingg et al., 1990). Dates from the lower crustal rocks are "very poorly constrained" (Downes, 1993). This paper discusses the isotopically dated two-stages of lower crust formation and the absence of the upper crust in the Anabar shield.

2. The Anabar shield: geological data

The Anabar shield is a rocky tundra plain forming a 63000 km^2 basement window through the Siberian platform between 69-72°N and 107-114°E (Fig.1). Major geological structures of the shield are well displayed on aerial photographs. The surrounding platform cover is comprised of Late Precambrian and Phanerozoic (mainly clastic) sediments. A K-Ar date of 1650 Ma obtained from the lowermost glauconite sandstones is interpreted as the age of deposition (Krylov et al., 1963). The shield is composed of three granulite terranes (80% of the shield surface) separated by two amphibolite grade shear zones (Rosen, 1989). Narrow isoclinal folds between 1-20 km wide and 10-50 km long in granulite

O. Oncken and C. Janssen (eds.), Basement Tectonics 11, 67–77.

terranes, roughly generalized on Fig.1, strike NNW and 60-80° dip to the ENE parallel to the main foliation in the shear zones. This orientation is assumed to be the result of thrusting to the WSW (Rosen et al., 1991, 1994).

The easternmost Hapschan terrane is composed of the Hapschan Group, which is largely made up of garnet bearing paragneisses, calc-silicates and marbles. Carbonate metasediments compose about 40% of the succession (Condie et al., 1991). Their Sm-Nd T(DM) model age of about 2400 Ma is interpreted by Zhuravlev and Rosen (1991) as a minimum age of an unknown eroded basement. Zircons from paragneisses yield an U-Pb date of 1970 Ma, interpreted as the last metamorphic event (Bibikova et al., 1988).

The central Daldyn terrane is predominantly metavolcanic two-pyroxene mafic to felsic granulites referred to as the Daldyn group. A Sm-Nd whole rock isochron suggests a protolith age of about 3100 Ma (Spiridonov et al., 1991) while U-Pb SHRIMP determinations on zircon indicate an age of about 3170 Ma. A multigrain discordia zircon date of 2760 Ma is thought to reflect the time of granulite facies metamorphism (Bibikova et al., 1988). The deepest horizon of the Daldyn terrane continental crust is probably represented by crustal xenoliths in the Udachnaya kimberlite pipe situated 200 km south of the Anabar shield. Mafic granulite xenoliths from this locality have tholeiitic compositions (Shatsky et al., 1990) and radioactive heat production as low as 0.08 $\mu W/m^3$ (Rosen,1992a). The Sm-Nd model age of these xenoliths is T(DM)=~3.0 Ga with a mineral isochron date of 1884 Ma. The latter is interpreted as the age of the youngest metamorphic event (Neymark et al., 1992).

The 10-20 km wide Billyakh shear zone separates the Hapschan and Daldyn terranes. The zone is composed of an amphibolite grade assemblage of mylonites, migmatites, large granite lenses and tectonic slices from the two adjacent terranes.

The westernmost Magan terrane includes the Upper Anabar Group, which is composed of highly deformed enderbites and charnockites with mafic to intermediate granulites. In places metacarbonate-paragneiss assemblages in this terrane are present in the folds. Mafic xenoliths from the Mirnyy kimberlite pipe, localized in this terrane 300 km south of the shield, have a Sm-Nd T(DM) model date of about 3020 Ma as a minimum age of the protolith. A mineral isochron of 1884 Ma probably reflects the age of the last metamorphic event (Neymark et al., 1992).

The Kotuykan shear zone, which separates the Magan and Daldyn terranes, is one of the largest tectonic boundaries in the Siberian plate (Rosen et al., 1994). It ranges in width from 10 to 30 km and is composed of anorthosite massifs and tectonic slices of granulites from the nearby

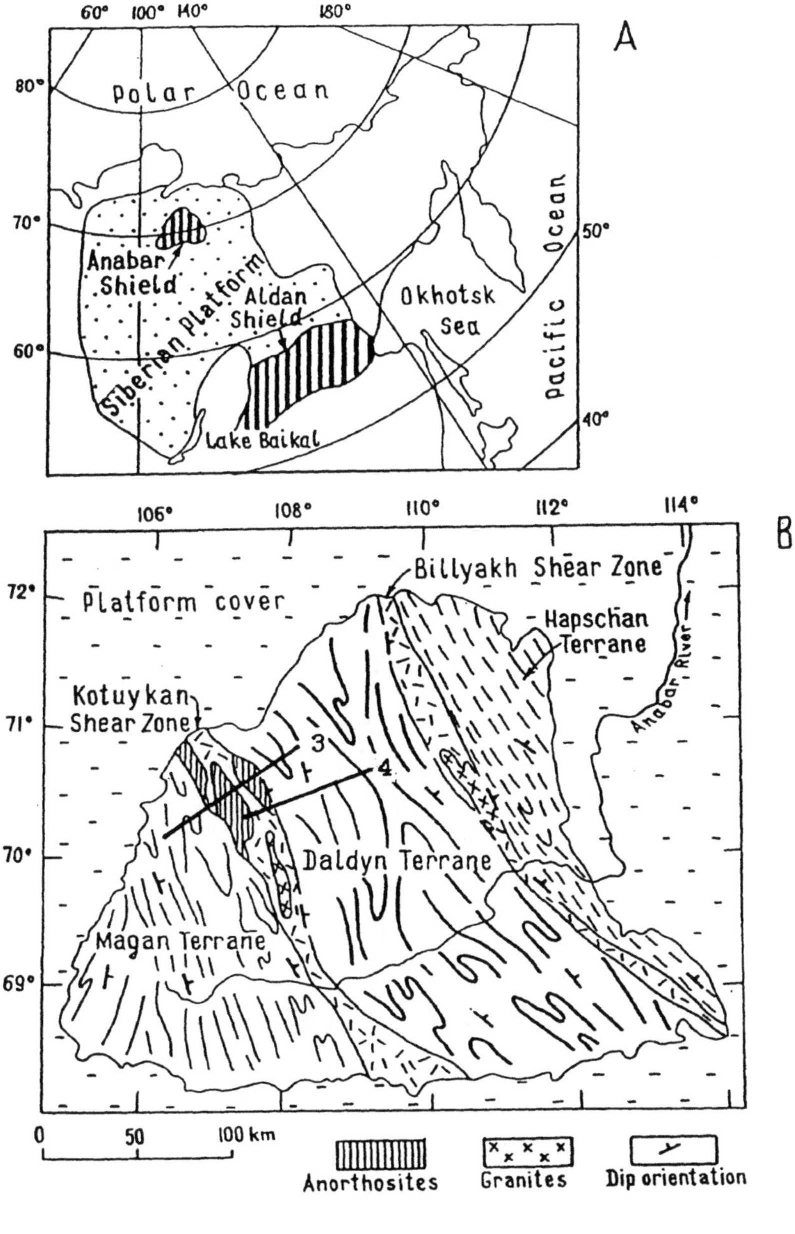

Figure 1. Geographical position (A) and geological sketch map (B) of the Anabar shield.
Lines 3 and 4 show positions of profiles, Fig.3 and Fig.4.

terranes. These tectonic slices and massifs are found as tectonic enclaves in gneisses, which are accompanied by granites and granodiorites, amphibolite grade mylonites and tectonized migmatites. P-T conditions of the migmatites and tectonites vary from amphibolite to greenschist facies (Fig. 2). The anorthosite and granulite enclaves have udergone retrograde metamorphism to amphibolite facies. The depth of the rock assemblage was calculated from negative gravity anomaly to be about 15 km (Fig. 3) and the exposed

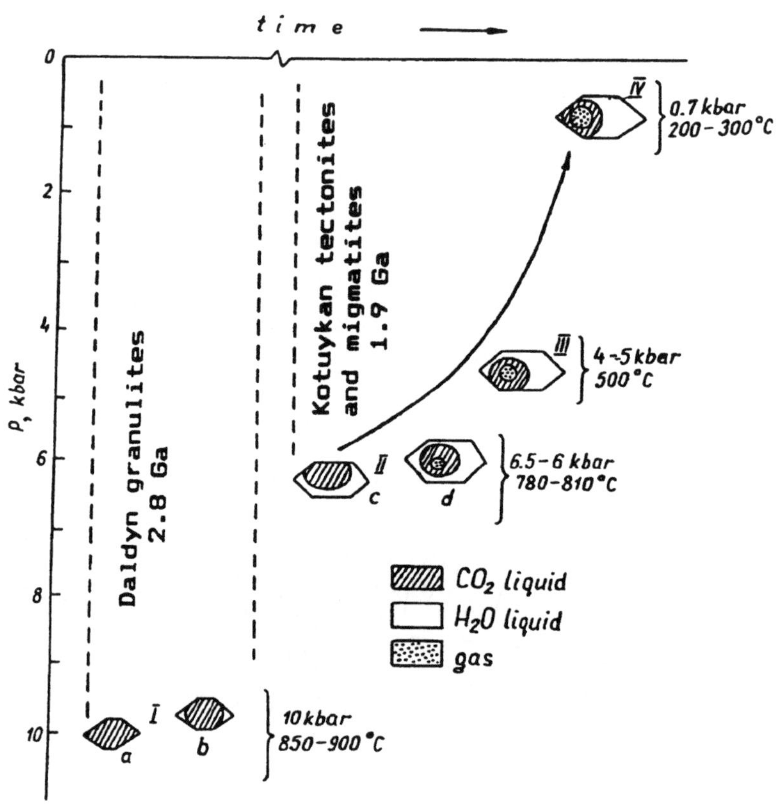

Figure 2. P-T parameters and compositions of fluid inclusions in quartz of metramorphic rocks in the Anabar shield (Rosen et al., 1991)

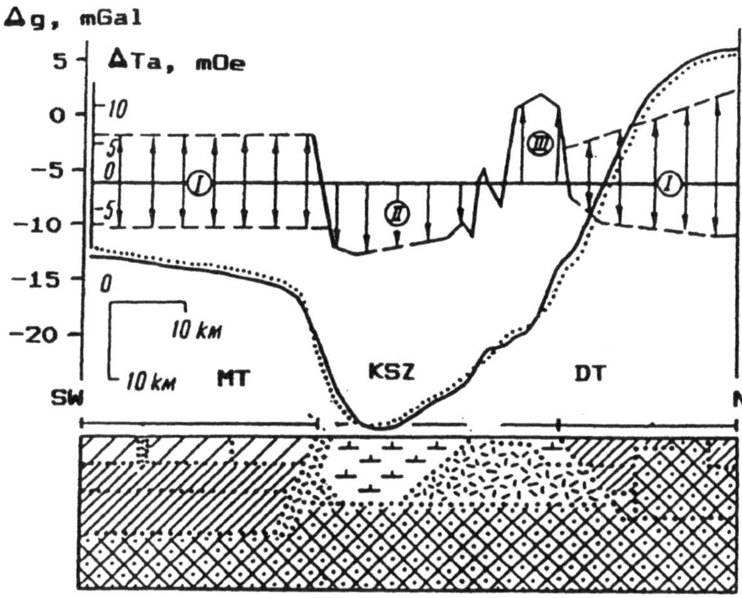

Kotuykan tectonites and migmatites, estimated density 2.6-2.72 g/cm^3

Anorthosites, 2.67 g/cm^3 and gabbro, 2.82-2.86 g/cm^3

Carbonates and gneisses, 2.73-2.76 g/cm^3

Orthopyroxene-plagioclase gneisses, 2.78-2.82 g/cm^3

Two pyroxene metabasites and orthopyroxene-plagioclase gneisses, 2.82-2,86 g/cm^3

Figure 3. Schematic gravity-magnetic and geological (calculated) profiles across the northern Kotuykan shear zone (Rosen et al., 1990) Abbreviations: MT - the Magan terrane, KSZ - the Kotuykan shear zone, DT - the Daldyn terrane. Figures in circles: intervals of magnetic anomalies of metamorphic rocks (I), anorthosites (II) and gabbros (III).

part of the Kotuykan shear zone apparently is a tectonic pile in front of upthrusted Daldyn terrane (Fig. 4). Migmatites in this zone yielded a U-Pb zircon age of 1925 Ma and granites a date of 1860 Ma (Rosen et al., 1991).

3. The Anabar shield: geophysical data

The low heat production (A=0.41-0.56 $\mu W/m^3$) calculated from radioactive element contents, rather high P-wave velocities (6.3-6.6 km/s), and rock densities (2.76-2.83 g/cm^3) are consistent with the interpretation that the granulite grade terranes outlined above are exhumed fragments of the lower crust, whereas the shear zones (with

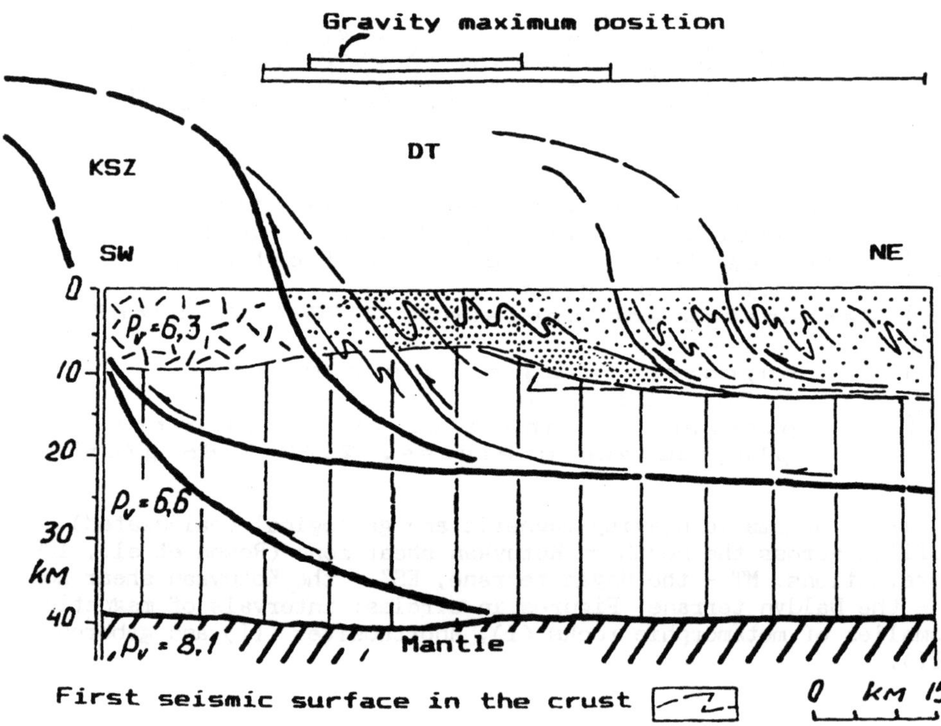

Figure 4. Schematic geological profile across the western front of the Daldyn terrane (DT) at the northern Kotuykan shear zone (KSZ). Dense dots denote mafic granulites, folded structure is generalized from geological mapping at a scale of 1:200,000.

values 1.22 μW/m³; 6.0-6.3 km/s and 2.72 g/cm³) represent the roots of more recently formed upper crust (Table 1). Rosen (1992a) concluded that the shear zones resulted from the collision of the Hapschan, Daldyn and Magan terranes. The generalized seismic profile shows the layered structure with several main layers: the uppermost layer (0-10 km deph) has a P_v=6.3 km/s, a density of 2.76 g/cm³ and radioactive heat production of A=0.76 μW/m³. The average wave velocities between 10-40 km are P_v=6.6 (6.3-7.6) km/s. The density increases from 2.83 g/cm³ in the 10-20 km layer up to 3.2 g/cm³ in the lowermost, 30-40 km, layer. Heat production A-values range from 0.41 μW/m³ to 0.08 μW/m³ (Rosen, 1992a). On average, the radioactive heat production is 0.36 μW/m³ for the whole crust which results in a crustal heat flow of 14.2 mW/m². This is 40% lower than the surface heat flow (25 mW/m²) measured on the Anabar shield, constraining models of its crustal composition and evolution. The geological and geochronological data above reflect the complicated structure and evolution of the lower continental crust exposed in the Anabar shield.

TABLE 1. Generalized metamorphic parameters and averaged physical properties of the Anabar shield geological subdivisions (weighted according to the rock volumes) (Rosen,1992a,b)

Geological units	P kbar	T °C	Density g/cm³	A μW/m³
Hapschan Group	5.5-7.0	750-820	2.74	1.18
Daldyn Group	8.5-11.0	820-950	2.83	0.41
Upper Anabar Gr.	7.0-8.5	700-850	2.76	0.56
Billyakh and Kotuykan shear zones	0.7-7.0	270-700	2.72	1.22
Shield whole surface	-	-	2.76	0.76

4.Discussion

Generalized P-T parameters in Table 1 are assumed to be the result of crustal thickening during the Archean and Early Proterozoic. Burial of the protoliths was associated with the formation of the first sialic blocks (microcontinents ?) in the Archean. The rocks of the Daldyn terrane exposed at the surface today were metamorphosed at a depth of 30 km under a low thermal gradient of 23-26 °C/km while those of the Magan terrane were formed at a depth of 25 km under a thermal gradient of 28-32 °C/km. The Archean tectonic environment of the shield is unknown, but low viscosity rocks under granulite grade metamorphic conditions and partial melting probably facilitated the development of recumbent isoclinal folding such as in the granulites of Granulitgebirge, Europe, and Antarctica (Sandiford, 1989). The Sm-Nd age of the lowermost horizons of the lower crust represented by mafic granulite xenoliths in the Siberian kimberlite pipes (Neymark et al.,1992) mentioned above is inconsisitent with a magmatic underplating process assumed to be widespread at the base of a pre-existing crust (Downes, 1993).

The Early Proterozoic collision (1.9 Ga) along the shear zones resulted in crustal thickening and mountain

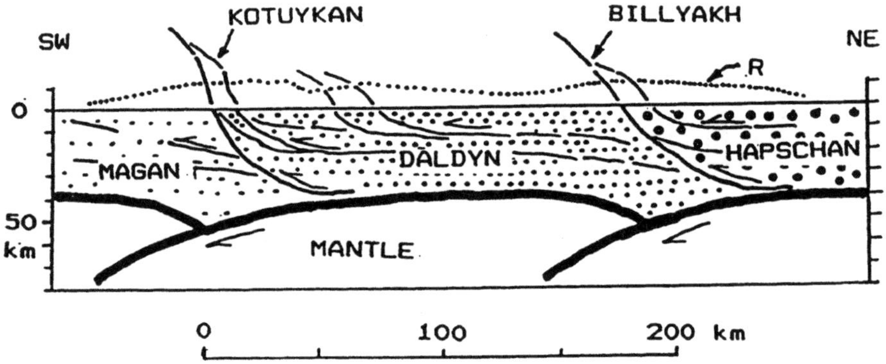

Figure 5. Paleo-geodynamic profile across earth's crust in the Anabar shield during the last stages of the Early Proterozoic collision.
R - assumed paleo-surface relief.

building, followed by erosion and exhumation of the lower continental crust at approximately 1650 Ma during the early Late Proterozoic (Fig.5). Metamorphic mineral assemblages of the rocks in the Hapschan terrane and in the Bilyakh and Kotuykan shear zones were formed at a maximum depth of 25 km under a high thermal gradient of 32-35 °C/km. The latest hydrothermal activity in these shear zones took place at a depth of 2-4 km. The uplift of 25 km to 2 km may be the result of erosion after collision.

At deep levels of the crust the Proterozoic collision resulted in the development of the subhorizontal tectonic banding in the granulite terranes associatd with decollements which evolved into listric thrust faults. The Archean recumbent folds in the shield were rotated into a steeply dipping series of isoclinal folds during this collision. Rosen et al., (1991) suggest that the main petrological events in the shear zones are: (1) retrograde metamorphism of granulites to amphibolite facies; (2) granite melt formation in shallow thrust-faults (decollements); (3) granite melt injected along the listric thrusts associated with the shear zones. The granite melts are thought to be the result of water-rich fluid penetrating along faults accompanied by an increase in temperature to the granite liquidus in the upthrusted plates (England and Thompson, 1984). The granitoids in the collision zones are, therefore, interprted to be relics of the granite-forming reservoirs within the thrust faults.

5. Conclusions

The Anabar shield undervent two stages of crust thickening.

1) Burial of an original Archean supracrustal pile during late Archean tectonic movements (collision ?). This formed the Archean lower crust at 2.8 Ga in the Daldyn and Magan terranes, two blocks were probably separated from each other before 1.9 Ga.

2) The Daldyn and Magan blocks and the Hapschan terrane collided at about 1.9 Ga. The resulting mountain building process was accompanied by the formation of granite melt and its intrusion into the upper crust probably to form granite-migmatite domes. By 200 Ma later these mountains were completely eroded and developments of the Siberian platform began with the deposition of Late Precambrian clastic cover sediments up to 2-5 km thick.

Two types of structures were present in the Anabar shield: 1) Isoclinal folds in the granulitic lower crust, and 2) Granitic domes in the amphibolite- and greenschist-grade upper crustal rocks.

6. Acknowledgments

This study was supported by the Anabar Project (1980-1990) of the Institute of the Lithosphere, Russian Academy of Sciences while the paper was prepared with the help of the International Basement Tectonics Association and the GeoForschungsZentrum Potsdam, Germany. The author thanks E.V.Bibikova, S.Yu. Milanovsky, V.L. Zlobin and D.Z. Zhuravlev for cooperation and D. Bridgwater, K.C. Condie, R.L. Rudnick and B.F. Windley for fruitful discussions on the geological and geochemical problems of the Anabar shield.

References

Bibikova, E.V., Belov, A.N. and Rosen O.M. (1988). Isotopic age of the Anabar complex and peculiarities of dating of the oldest geological events, *in* M.S. Markov (ed), *The Archean of the Anabar Shield and the problems of early evolution of the Earth*, Nauka Press, Moscow, pp.115-133 (in Russian).

Condie, K.C., Wilks, M., Rosen O.M. and Zlobin, V.L. (1991) Geochemistry of metasediments from Precambrian Hapschan series, eastern Anabar shield, Siberia, *Precamb.Research* **50**, 37-47.

Downes, H.(1993) The nature of the lower continental crust of Europe: petrological and geochemical evidence from xenoliths, *Physics of the Earth and Planet. Inter.*,**79**, 195-218.

England P.C., Thompson A.(1984) Pressure-temperature-time paths of regional metamorphism: 1.Heat transfer during the evolution of regions of thickened continental crust, *Jour.of Petrology* **25:4**, 894-928.

Krementtsky, A.A., and Milanovsky S.Yu.(1986) Geothermal study and the heat production model of the Precambrian crust in the north-eastern Baltic Shield, *in* L.N. Ovchinnikov (ed), *Geochemistry of the deep-seated rocks*, Nauka Press, Moscow, pp.131-149 (in Russian).

Krylov, A.Ya., Vishnevsky, A.N. and Silin, B.I.(1963) Absolute age of rocks in the Anabar shield. *Geokhimia*, **12**, 1140-1144 (in Russian).

Meissner R.(1986) *The continental crust*, Academic Press Inc., Orlando-San Diego-N.Y.-Toronto, 402 p.

Neymark, L.A., Nemchin, A.A., Rosen, O.M.,Serenko, V.P., Spetsius, Z.V. and Shuleshko, I.K.(1992) Sm-Nd isotopic systems in the lower crust xenoliths from kimberlites of Yakutia. *Doklady, Russian Acad. Sci.* **327:3**, 374-378 (in Russian).

Percival, J.A., Fountain, D,M., Salisbury,M.H.(1992) Exposed crustal sections as windows on the lower crust,*in* D.M. Fountain, R.J. Arculus and R.W. Kay (eds.), *Continental lower crust*, Elsevier, Amsterdam, pp.317-362.

Rosen O.M.(1989) Two geochemically different types of Precambrian crust in the Anabar shield, north Siberia. *Precamb. Research* **45**, 129-142.

Rosen, O.M.(1992a) Heat production of the Anabar shield earth crust and problems of the lower continental crust formation, *Geol.i Geofisika*, **12**, 22-29 (in Russian).

Rosen, O.M.(1992b) Geochemistry of granulites and the Archean sialic crust formation in the Anabar Shield, North Siberia, *in* S.S.Augustithis (ed), *High grade metamorphics*, Teophrastus Publications S.A., Athens, Greece, pp.69-102.

Rosen, O.M., Condie, K.C., Natapov, L.M. and Nozhkin, A.D.(1994) Archean and Early Proterozoic evolution of the Sberian craton: a preliminary assessment, *in* K.C. Condie (ed), *Archean crustal evolution*, Elsevier, Amsterdam, pp.411-459.

Rosen, O.M., Rachkov, V.S. and Sonyushkin, V.E.(1991) Granitization of tectonites and granite formation in shear zones of the Anabar shield, *Geokhimia* 9, 1240-1249 (in Russian).

Rosen, O.M., Zlobin, V.L. and Rachkov, V.S.(1990). The Anabar Shield, *in* Yu.M. Pushcharovsky and V.G. Trifonov (eds), *Lithosphere tectonic layering and regional gelogical studies,* Nauka Press, Moscow, pp. 269-280 (in Russian).

Rudnick, R.L.(1992) Xenoliths - Samples of the lower continental crust, *in* D.M. Fountain, R.J., Arculus and R.W. Kay (eds.), *Continental lower crust,* Elsevier, Amsterdam, pp.269-317.

Rutter, E.H. and Brodie, K.H.(1992) Rheology of the lower crust, *in* D.M. Fountain, R.J. Arculus and R.W. Kay (eds.), *Continental lower crust,* Elsevier, Amsterdam, pp.201-268.

Sandiford, M.(1989) Horizontal structures in granulite terranes: a record of mountain building or mountain collapse? *Geology,* **17**, 449-452.

Shatsky,V.S., Rudnick,R.L. and Jagoutz,E.(1990) Mafic granulite xenoliths from Udachnaya pipe, Yakutia: samples of Archean lower crust?, *in* N.V. Sobolev (ed.), *Deep seated magmatism and evolution of lithosphere of the Siberian platform.* IGG SO AN USSR, Novosibirsk, pp.35-38.

Spiridonov, V.G., Sukhanov, M.K., Karpenko, S.F. and Lyalikov, A.V.(1991) Sm-Nd isotope system of the oldest granulites in the Anabar shield, *Doklady, Russian Acad. Sci.* **319**:5, 1209-1212 (in Russian).

Taylor, S.R. and McLennan,S.(1985) *The continental crust: its composition and evolution (an exhumation of the geochemical record preserved in sedimentary rocks),* Blackwell, Oxford, 312 p.

Zhuravlev, D.Z. and Rosen O.M.(1991) Sm-Nd model age of metasediments in the Anabar shield granulite complex. *Doklady, Russian Acad. Sci.* **317**:1, 189-193 (in Russian).

Zingg,A., Handy,M., Hunziker,J. and Schmid,S.M.(1990) Tectonometamorphic history of the Ivrea zone and its relationship to the crustal evolution of the southern Alps. *Tectonophysics,* **182**, 169-182.

STRUCTURAL SETTING OF THE PALEOZOIC CRYSTALLINE BASEMENT OF THE NORTHERN APENNINES (ITALY)

F.M. ELTER[1]* and E. PANDELI[2]
[1]*Department of Earth Sciences, University of Siena, Via delle Cerchia 3, 53100 Siena, Italy*
[2]*Department of Earth Sciences, University of Florence, Via G. La Pira 4, 50121 Florence, Italy*

ABSTRACT

The upper part of the crystalline basement of the Northern Apennines is made up of a complex pile of tectonic wedges derived from Alpine low-grade metamorphic units (T= temperature = 300°-450°C; P= pressure = 3-4 Kbar), Paleozoic (?Late Cambrian-Permian) to Tertiary in age. The Paleozoic rocks also show evidence of pre-Alpine deformation (e.g. Hercynian schistosity relics and folding).

In the subsurface of the Larderello geothermal region (southern Tuscany) this tectonic wedge complex (Monticiano-Roccastrada Unit) also includes Hercynian low-/medium-grade metamorphic rocks (Micaschist Complex) which are strongly imprinted by the Alpine tectono-metamorphism. At a depth of about 2000 m (below the ground level), an Alpine mylonite horizon divides the Micaschist Complex from an underlying highly recrystallized succession (Gneiss Complex) which shows a well preserved Hercynian framework modified only by weak Alpine crenulations. The Hercynian near-isothermal decompressional tectono-metamorphic evolution of the Gneiss Complex is similar to that of the Micaschist Complex and matches those of several areas of the Variscan basement in SW Europe with a barrovian-type intermediate P event (T= 550°-650°C; P≅ 6-7Kbar) to a syn-tectonic high T-low P event (T= 500°-600°C; P= 2.8-3.5 Kbar).

The tectonic superposition of the Alpine units (including the Micaschist Complex) on the Hercynian-preserved Gneiss Complex could be the result of their being overthrust onto the undeformed basement of the Apenninic foreland or onto "rigid" Hercynian section belonging to the Northern Apennines pile.

Geophysical studies reveal probable deeper crustal shear-zones in the inner part of the mountain chain (3-8 Km deep seismic "K" horizon; 10-14 Km deep doubling of the crystalline basement in the magnetic and gravimetric profiles). Collectively, these data point to a strong Alpine reactivation of the Hercynian basement of the Northern Apennines at relatively deep crustal levels.

1. Introduction

The metamorphic Paleozoic rocks of the Northern Apennines are part of the southern segment of the Hercynian chain (Vai, 1978; Bagnoli et al., 1979; Vai and Cocozza, 1986; Sassi and Zanferrari eds.,1989; Pandeli et al., 1994 cum bib.) which was strongly reactivated during the Alpine Orogeny.

In the last few years, many advances were made in the knowledge of the stratigraphy and of the Alpine and pre-Alpine structural evolution of these metamorphic sequences (Carmignani and Kligfield, 1990; Bertini et al., 1991; Conti et al., 1991 a, b; Elter and Pandeli, 1990,1991, 1993; Pandeli et al., 1991, 1994).

This paper presents an up-to-date view on the stratigraphy and on the structural features of the Paleozoic sequence, which represent an important "key" in the reconstruction of the tectonic setting of the Northern Apennines crystalline basement.

O. Oncken and C. Janssen (eds.), Basement Tectonics 11, 79–90.

2. Geological outline

The Northern Apennines is a part of the Alpine belt which is characterized by the Adria-vergent piling of units derived from the western Tethys oceanic domain (Ligurids) onto those of the Adriatic continental margin (Tuscan-Umbrian units) during the Paleogene-Neogene (Sestini ed., 1970; Giannini and Lazzarotto, 1975; Kligfield, 1979; Boccaletti et al., 1980; Carmignani and Kligfield, 1990).

In this complex tectonic pile, the metamorphic successions (Tuscan metamorphic units = TMU) are overlain by the anchimetamorphic to diagenetic Mesozoic-Tertiary formations of the Tuscan-Umbrian (e.g. the Tuscan Nappe) and of the Ligurid sedimentary units. The TMU is made up of a basal Paleozoic section, which represent the deepest crustal rocks cropping out in the Northern Apennines (see Fig.1), that is overlain unconformably by Triassic-Oligocene "cover" formations (e.g. in the Apuan Alps area). These latter include Triassic continental-coastal metasiliciclastics ("Verrucano"), Triassic-Liassic shallow marine carbonate and Liassic-Oligocene pelagic carbonate-siliceous-terrigenous metasediments.

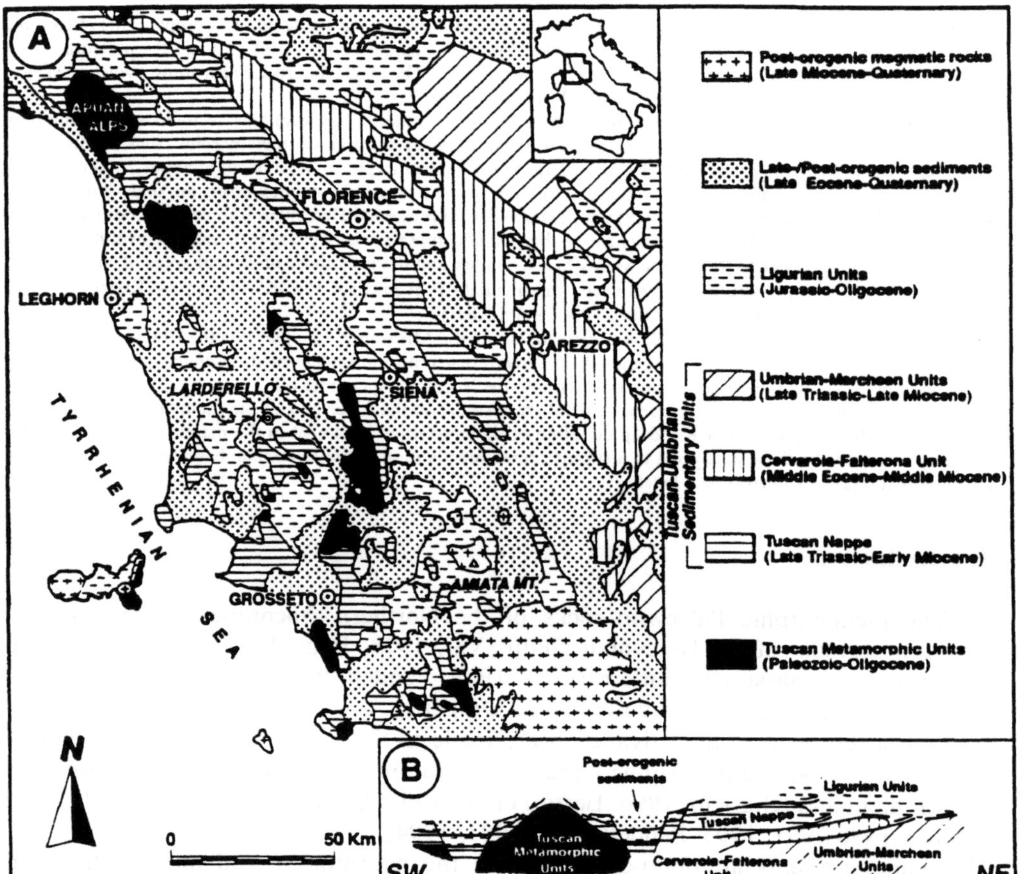

Figure 1. A) Geological sketch map of the Northern Apennines; B) Schematic cross-section.

The TMU are characterized by polyphased low-grade regional metamorphism, in which temperature ranged between 300°-450°C and pressure between 3-4 Kbar (Di Pisa et al., 1985; Franceschelli et al., 1986). Oligocene-Miocene (27→12 Ma) K/Ar and $^{40}Ar/^{39}Ar$ dates on the Apuan Alps rocks (Kligfield et al., 1986) apparently reflect the time of metamorphism. The Alpine tectogenetic events also caused complex wedging and piling of the units (Pandeli et al., 1991).

On a regional scale, the TMU are exposed along the Tuscan mid-ridge (from the Apuan Alps to Grosseto) and in other scattered outcrops in the inner part of the chain (see Fig. 1). They have also be found in several deep wells in the Larderello and Amiata Mt. geothermal fields and in the Massa Marittima mining region (near the Larderello field) of southern Tuscany.

3. The Paleozoic sequences of the Northern Apennines

3.1 STRATIGRAPHY

The strong overprint of the Alpine orogeny obliterated most of the original textures and fossils. Fossils were found only in the Silurian-Devonian and in the Carboniferous-Permian strata. The comparison of the Tuscan Paleozoic formations with similar, well-dated Late Cambrian to Carboniferous successions of central Sardinia give us a key in the stratigraphic interpretation of the unfossiliferous pre-Triassic rocks of the Northern Apennines.

The restored Tuscan Paleozoic "type"-sequence (Pandeli et al., 1994) is made up of four main sections (Fig. 2). From the bottom they are as follows:

1) ?Late Cambrian-?Early Ordovician basinal metasiliciclastics with intercalated metabasites. The Sardinian (eo-Caledonian) deformation event caused the emersion of these rocks.

2) ?Middle Ordovician-Devonian transgressive section: a) sub-aerial acidic to intermediate metavolcanics and metasediments (?Middle Ordovician); b) coastal-marine metasiliciclastics (?Late Ordovician); c) neritic-pelagic anoxic (Silurian-Early Devonian) to well oxygenated (Middle-Late Devonian) metapelites, metacarbonates and minor quartzites. During the Middle-Late Ordovician some metabasitic dykes were probably emplaced. In southern Tuscany the angular unconformity between the Devonian rocks and the overlying ?Early Carboniferous fluvial-deltaic metasediments reflects to the onset of the Hercynian orogeny (Bretonian phase). Pre-Alpine schistosity relics in the rocks older than Late Visean are thought to be the result of an Early Carboniferous tectono-metamorphic event (Sudetian event).

3) Early (Late Visean)-Middle Carboniferous Hercynian foredeep sediments: pre-flysch silty-shaly sediments passing upward to arenaceous turbidites with local intercalated calcarenites and carbonate olistoliths. At the end of the Middle Carboniferous the Asturian folding event took place.

4) Late Carboniferous-Permian "molasse"-type continental to marine sediments, which were deposited in late-/post-orogenic extensional basins. Acidic volcanics occur locally in the Permian sequence.

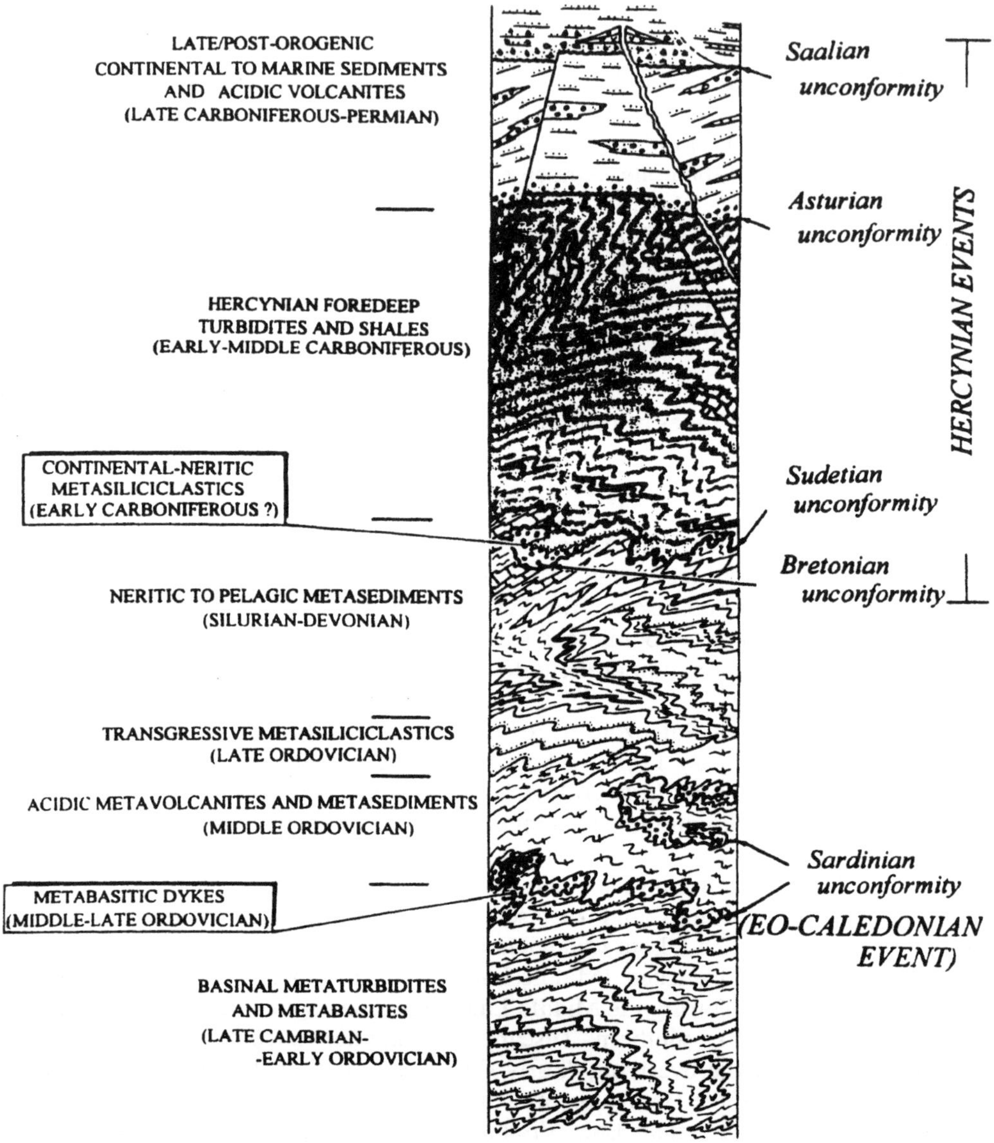

Figure 2. Restored Paleozoic "type"-sequence of the Northern Apennines basement.

3.2 STRUCTURAL-METAMORPHIC FEATURES

The Paleozoic sections of the TMU, as well as their Triassic to Tertiary cover rocks, show evidence of three Alpine deformation events (D_1, D_2 and D_3), the first two accompanied by greenschist facies metamorphism (T=300°-450°C; P=3-4 Kbar).

The structural-metamorphic framework of the Alpine events in the TMU of southern Tuscany (Elter and Pandeli, 1990, 1993) is summarized below :

1) D_1 is mainly represented by NW/SE-trending and east-facing near isoclinal folds with an associated penetrative schistosity (S1=sericite+quartz+chlorite+opaques± albite). The angular relationships between the ellipsoid axis, the S_1 foliation and the ellipsoid shape suggest that D_1 developed from simple shear (Elter and Meccheri, 1994).

Between the D_1 and D_2 events, a peculiar static blastesis of albite, chloritoid and kyanite (only in the Verrucano successions) took place.

2) D_2 is generally characterized by close to tight folds with gradational to discrete crenulation cleavage (*Domain I*). Data from wells in the Larderello geothermal field indicate that the D_2 structural framework changes to tight to isoclinal, at times sheath-type (Elter, 1993) folds, with a spaced penetrative schistosity (S_2= sericite± quartz±chlorite±opaques) (*Domain II*), at deeper structural levels.

The D_2 folds are largely coaxial with the D_1 folds, and generally verge toward SW.

3) D_3 is defined by open, generally NE/SW-trending, polyclinal folds with poorly developed crenulation cleavage.

This Alpine tectono-metamorphic framework in southern Tuscany is similar to that described in the literature for the TMU of northern Tuscany (e.g. Apuan Alps: Carmignani et al., 1978, 1980; Boccaletti et al., 1983; Kligfield et al., 1986; Moratti et al., 1989; Carmignani and Kligfield, 1990).

Deformation D_1 was the main tectonic event in the Northern Apennines, while D_2 and D_3 were later extensional events resulting from exhumation of the overthickened tectonic pile (Carmignani and Kligfield, 1990).

However, the post-tectonic static blastesis of HT-LP (high temperature-low pressure) minerals (biotite, andalusite, cordierite, corundum, garnet, wollastonite, etc.), associated with the Late Miocene-Quaternary anatectic magmatism, occurred only in the Paleozoic-Mesozoic rocks of southern Tuscany.

The Hercynian structural evolution of the Tuscan Paleozoic sequences (Conti et al., 1991 a,b; Elter and Pandeli, 1990; Pandeli et al., 1994) is mainly reflected in the schistosity relics (muscovite+quartz+chlorite±albite±opaques) of the Sudetian event in the pre-Late Visean rocks. Moreover, evidence of the pre-Triassic folding of the Hercynian flysch (Asturian event) and the later rifting events (e.g. Saalian event) is locally present in the Paleozoic sequences of Tuscany (Bagnoli et al., 1979; Costantini et al., 1987/88; Conti et al., 1991a; Pandeli et al., 1994) (see Fig. 2).

4. Deep wells of the Larderello geothermal region

The deep (4000-4500 m below the ground level) wells of the Larderello geothermal region allow us to reconstruct a complete upper crustal section of the Northern Apennines basement (Elter and Pandeli, 1990).

Below the Ligurids and the Tuscan Nappe, the wells cross the following tectonic units (Fig. 3):

1) *Monticiano-Roccastrada Unit*. The upper part of this unit is a complex pile of tectonic wedges made up of Early to Late Paleozoic formations, and Triassic "Verrucano" and Late Triassic anhydrite-carbonate strata. The lower part is made up of

?Late Cambrian-?Early Ordovician siliciclastics and metabasites (Phyllitic-quartzitic Complex) and of the underlying, possibly ?early Paleozoic-?Precambrian, Micaschist Complex.

2) *Gneiss Complex.* It is composed of polymetamorphic rocks of unknown age (?early Paleozoic-?Precambrian). Late Alpine NW/SE-trending folds, similar to a "gneiss dome" (Bertini et al., 1994), are present in this tectonic unit.

The contact between the Micaschist Complex and the Gneiss Complex (at a depth of about 2000 m below the ground level) is marked by Alpine mylonites.

4.1 THE MICASCHIST COMPLEX

The Micaschist Complex, which has a highly variable thickness (from 100 to 900 m), consists of low-/medium-grade, garnet-bearing micaschists with varying amounts of amphibolites. Although this unit has a strong Alpine imprint (structural features of the Domain II), its Hercynian structural-metamorphic evolution is still recognizable.

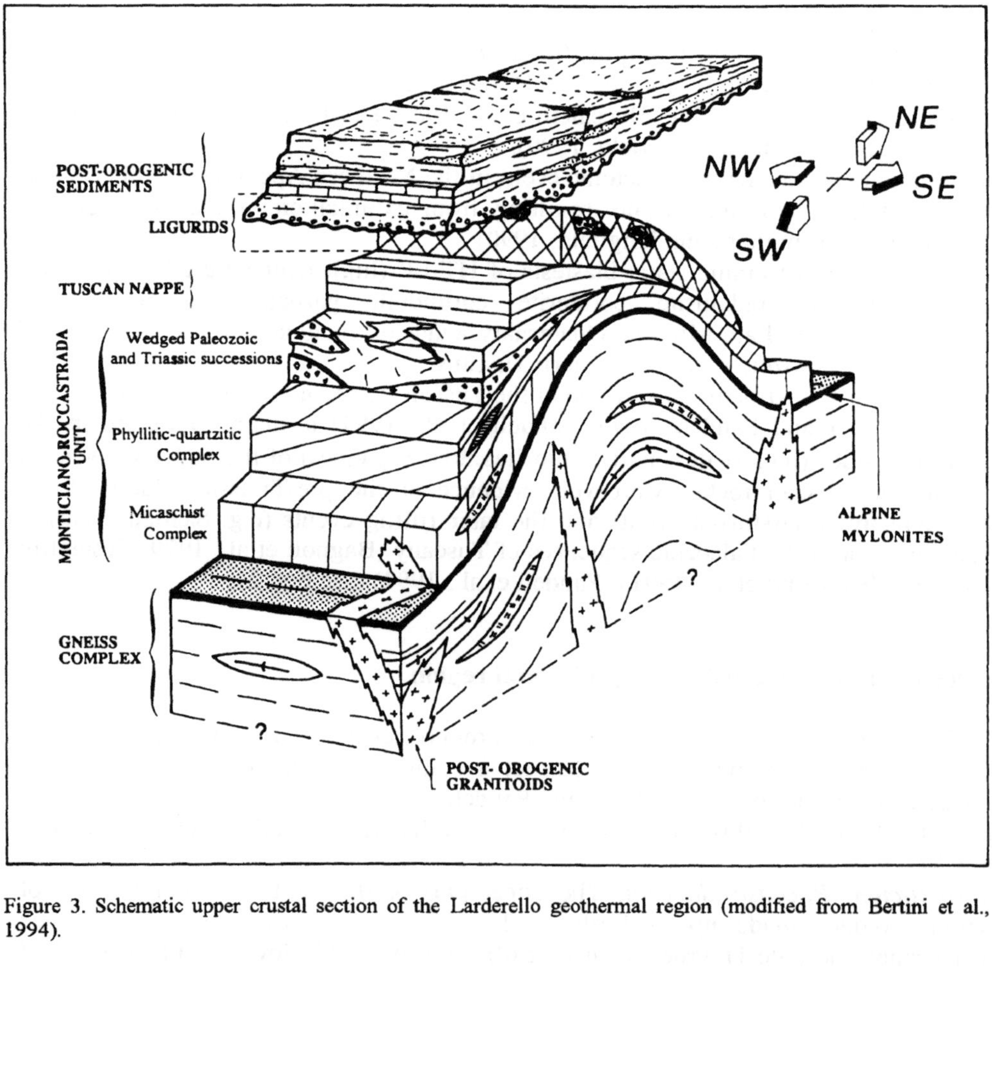

Figure 3. Schematic upper crustal section of the Larderello geothermal region (modified from Bertini et al., 1994).

The following sequence of Hercynian events has been delineated:
1) Initial crystallization of albite/oligoclase and garnet (core: Alm62-67, Pyr3-4, Grs16-19, Sps3-6; rim: Alm76-84, Pyr6-7, Grs3-16, Sps0-3) porphyroblasts, sometimes including S_x (=graphite±quartz±muscovite) trails.
2) Subsequent syn-tectonic blastesis of biotite+muscovite+quartz±albite/oligoclase, reflected in the Hercynian schistosity relics (S_e).
3) Later HT-LP crystallization of andalusite±cordierite±biotite±muscovite. Rb/Sr radiometric analyses on the muscovite indicate that this event occurred about 285 my ago (c.f. Del Moro et al., 1982).

4.2 THE GNEISS COMPLEX

The Gneiss Complex contains medium- to high-grade gneiss and paragneiss with varying amount of amphibolites, amphibolitic gneiss, granitic to granodioritic orthogneiss and augen-gneiss. Alpine quartz-mylonites are common at the top of this Complex.
The Gneiss Complex shows a well preserved Hercynian framework which was modified only by weak Alpine crenulations and by the post-tectonic contact blastesis.

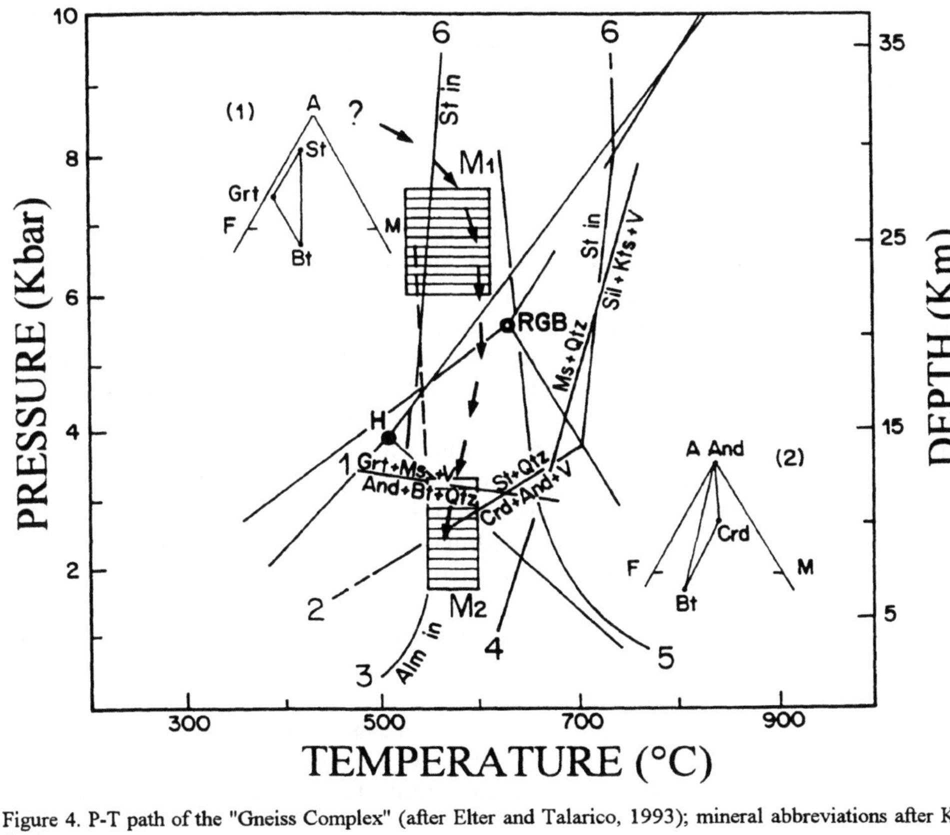

Figure 4. P-T path of the "Gneiss Complex" (after Elter and Talarico, 1993); mineral abbreviations after Kretz, 1983.

The pre-Alpine structural-metamorphic evolution of the Gneiss is similar to that of the Micaschist Complex (Elter and Pandeli, 1990; Elter and Talarico, 1993; Bertini et al., 1994; Elter and Pandeli, unpublished data), and its P-T trajectory path shows near isothermal decompression during the Hercynian orogeny (Fig. 4).
The following two metamorphic events have been identified in the complex:
1) An earlier barrovian-type intermediate P event (M_1 event in Fig. 4) with T= 550-650°C and P≅ 6-7Kbar (Elter and Talarico, 1993), that produced the crystallization of albite/oligoclase and garnet (cores: Alm73-82, Pyr6-10, Grs7-14, Sps1-7; rim: Alm78-83, Pyr5-7, Grs3-8, Sps7-9) porphyroblasts, ±staurolite (X_{Mg}= 0.12, ZnO = 1.75 wt%) ±prismatic sillimanite. Graphite-rich S_x inclusion trails are sometimes observed in the plagioclase and in the garnet porphyroblasts.
2) A subsequent syn-tectonic HT-LP event (M_2 event in Fig. 4) with T= 500-600°c and P= 2.8-3.5 Kbar (Elter and Talarico, 1993) that is reflected in the main pervasive composite schistosity (S_e) of these rocks. S_e is defined by the oriented blastesis of quartz+muscovite+biotite+andesine/oligoclase±hornblende, with younger cordierite (X_{Mg}= 0.50) and andalusite. The muscovite has a low-celadonite content with Si^{4+}= 3.0-3.17 a.f.u, Ti< 0.06 a.f.u. and Na/Na+K<0.13; biotite is characterized by an intracrystalline homogeneous composition with X_{Mg}= 0.35-0.38, Al^6= 0.50-0.75 a.f.u. and Ti< 0.3 a.f.u.
The earlier barrovian minerals were partly re-equilibrated during this event (ex. cordierite±andalusite±hercynite after staurolite; andalusite±biotite±quartz±cordierite± plagioclase coronas of the garnet).

5. Final remarks

On a regional scale, the sedimentary units (e.g. Ligurids and Tuscan Nappe) rest on a complex pile of tectonic wedges (the Monticiano-Roccastrada Unit of southern Tuscany), mostly made up of Paleozoic and Triassic low-grade metamorphic formations. In the subsurface of the Larderello area, the Monticiano-Roccastrada Unit also includes a low-/medium-grade metamorphic unit (Micaschist Complex). All these units show a strong Alpine polyphase low-grade tectono-metamorphic imprint (D_1, D_2 and D_3 events that occurred between 27-12 Ma). In the pre- Late Visean rocks the Alpine structures are clearly superimposed onto a previous tectono-metamorphic framework (e.g. the schistosity relics) that developed during the Hercynian orogeny.
These data confirm that the Paleozoic basement of the Northern Apennines (as well as the eastern and Southern Alps, Corsica-Sardinia and Calabrian-Peloritan pre-Alpine sequences) is part of the southern Hercynian chain and was strongly reactivated during the Alpine orogeny.
Additionally, borehole data from the deep geothermal wells of the Larderello region indicate that the Alpine units (including the Micaschist Complex) are tectonically superposed on a Hercynian sequence (Gneiss Complex).

The blastesis-deformation relationships of the Gneiss Complex (similar to the pre-Alpine ones of the Micaschist Complex) points to a continuous Hercynian tectono-metamorphic evolution from initial crustal thickening to post-thickening exhumation (Fig 4).

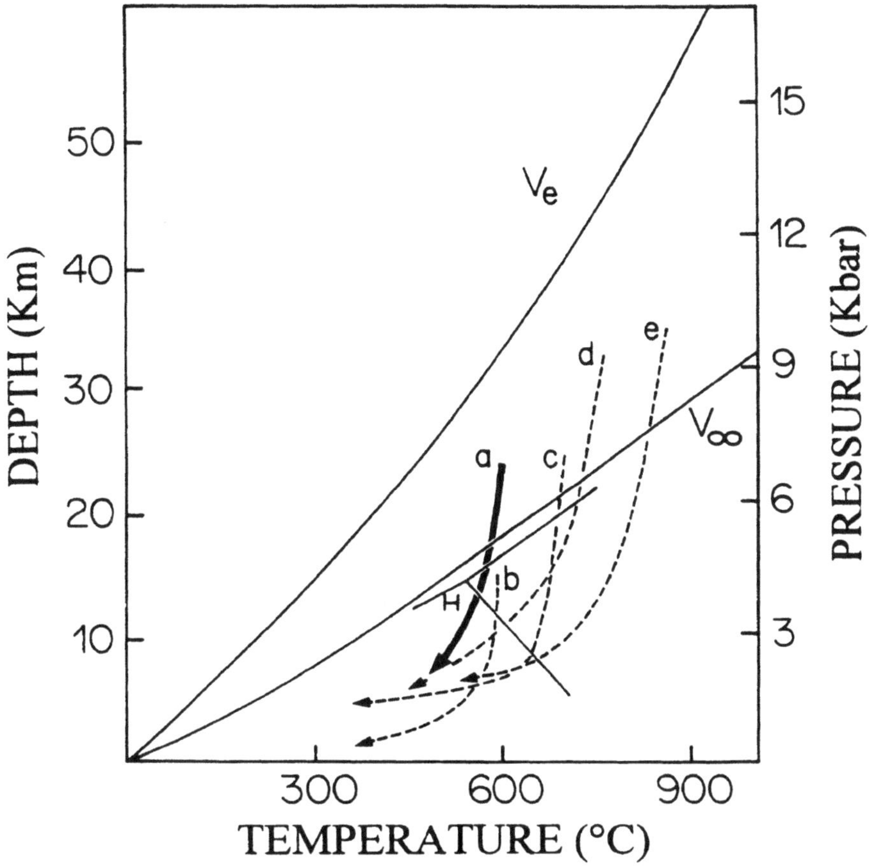

Figure 5. P-T paths of some Hercynian crystalline basements in southwestern Europe (after Elter and Talarico, 1993): a) "Gneiss Complex" (Larderello geothermal region, Italy); b) Mandatoriccio Complex (Calabrian-Peloritan arc, Italy); c) Migmatitic complex (NE Sardinia, Italy); d) Migmatitic complex (Central Massif, France); e) Granulitic complex (Bethic-Rifean belt, Spain).

This near-isothermal decompression path suggests an important role for extensional tectonism in the exhumation of the intermediate P metamorphic rocks in this part of the Hercynian orogenic belt. Similar P-T paths have been determined for other parts of the Hercynian basement of southwestern Europe: Central Massif (Mercier et al., 1991), NE Sardinia (Elter et al., 1986; Ricci, 1992; Di Pisa et al., 1993), western and Southern Alps (Dal Piaz, 1993), Calabrian-Peloritan arc (Borghi et al., 1992), Bethic-Rifean belt (Kornprobst and Vielzeuf, 1984) (Fig. 5). In this model the 285±11 Ma data, obtained for the late-tectonic muscovite of the Micaschist Complex, could represent a cooling age due to the pre-Permian exhumation or a late-/post-tectonic magmatic event such as the Late Carboniferous-Permian syn-tectonic magmatism of Sardinia (Ghezzo and Orsini, 1982).

The tectonic superposition of the Monticiano-Roccastrada Unit onto the Gneiss Complex may be the result of the overthrusting of the Alpine Units onto the undeformed basement of the Apenninic foreland (Bertini et al., 1991). An alternative hypothesis is that the Gneiss Complex is a "rigid" wedge of Hercynian rocks within the Northern Apennines tectonic pile. In this hypothesis, the HT-LP metamorphic anhydrite-dolostone horizon, recently found within the Gneiss Complex of the Larderello subsurface, could be attributed to a tectonic intercalation of Triassic evaporites into the Hercynian sequence.

In either case, it is clear that the Alpine orogeny affected deep crustal levels of the Northern Apennines crystalline basement.

The presence of Alpine crustal shear-zones within the crystalline basement of the Northern Apennines is also supported by geophysical data. These data are:

1) A 3-8 km deep main seismic reflecting horizon ("K" horizon) in southern Tuscany, probably related to an Alpine fractured shear horizon (Bertini et al., 1991; Cameli et al., 1993; Decandia et al., 1994).

2) Magnetic and gravimetric profiles which indicate a probable regional doubling of the acidic crystalline basement in the inner part of the Apennine chain at a depth of 10-14 km (Arisi Rota and Fichera, 1987; Cassano, 1991; Cassinis et al., 1991; Gualtieri et al.,1992).

Acknowledgements

The Authors wish to thank Prof. E. Abbate, Prof. A. Lazzarotto, Dr. M. Puxeddu and an anonymous referee for the critical reading and improving of the manuscript. Thanks are also extended to Dr. C.Verrucchi for some microprobe analyses and to Miss A. Mancini for the drafting of figures. This research was supported by C.N.R. (National Research Council of Italy) n. 90.00822 (Prof. A.Lazzarotto) and C.N.R. - Centro di Studio per la Geologia dell'Appennino e delle Catene perimediterranee (publ. n. 280).

References

Arisi Rota, F. and Fichera, R. (1987) Magnetic interpretation related to geo-magnetic provinces: the Italian case history, *Tectonophysics* **138**, 179-196.

Bagnoli, G., Gianelli, G., Puxeddu, M., Rau, A. and Tongiorgi, M. (1979) A tentative stratigraphic reconstruction of the Tuscan Paleozoic basement, *Mem. Soc. Geol. It.* **20**, 99-116.

Bertini, G., Cameli, G.M., Costantini, A., Decandia, F.A., Di Filippo, M., Dini, I., Elter, F.M., Lazzarotto, A., Liotta, D., Pandeli, E., Sandrelli, F. and Toro, B. (1991) Struttura geologica fra i Monti di Campiglia e Rapolano Terme (Toscana Meridionale): Stato attuale delle conoscenze e problematiche, *Studi Geologici Camerti*, Vol. Spec., 1991/1, 155-178.

Bertini, G., Elter, F.M. and Talarico, F. (1994) Evidenze di una fase estensionale pre-triassica nel complesso degli gneiss nell' area geotermica di Larderello (Toscana meridionale), *Studi Geologici. Camerti*, Vol. spec., 1994/1, 129-137.

Boccaletti, M., Capitani, S., Coli, M., Fornace, G., Gosso, G., Grandini, G., Milano, P.F., Moratti, G., Nafissi, P. and Sani, F. (1983) Caratteristiche deformative delle Alpi Apuane settentrionali, *Mem Soc. Geol. It.* **26**, 527-534.

Boccaletti, M., Coli, M., Decandia, F.A., Giannini, E. and Lazzarotto, A. (1980) Evoluzione dell'Appennino Settentrionale secondo un nuovo modello strutturale, *Mem. Soc. Geol. It.* **21**, 359-373.

Borghi, A., Colonna, V. and Compagnoni, R. (1992) Structural and metamorphic evolution of the Bocchigliero and the Mandatoriccio complexes in the Sila Nappe (Calabrian-Peloritan arc, southern Italy), in L.Carmignani and F.P. Sassi (eds.), *Contribution to the geology of Italy with special regard to the Paleozoic basements - A volume dedicated to Tommaso Cocozza*, IGCP N.276, *Newsletter* **5**, 321-334.

Cameli, G.M., Dini, I. and Liotta, D. (1993) Upper crustal structure of the Larderello geothermal field as a feature of post-collisional extensional tectonics (southern Tuscany, Italy), *Tectonophysics* **224**, 413-423.

Carmignani, L. and Kligfield, R. (1990) Crustal extension in the northern Apennines: the transition from compression to extension in the Alpi Apuane core complex, *Tectonics* **9**, 1275-1303.

Carmignani, L., Giglia, G. and Kligfield, R. (1978) Structural evolution of the Apuan Alps: An example of continental margin deformation, *Journ. of Geology* **86**, 487-504

Carmignani, L., Giglia, G. and Kligfield, R. (1980) Nuovi dati sulla zona di taglio ensialica delle Alpi Apuane, *Mem. Soc. Geol. It.* **21**, 93-100.

Cassano, E. (1991) Dati magnetici lungo il profilo CROP 03, *Studi Geologici Camerti*, Vol. spec., 1991/**1**, 49-53.

Cassinis, R., Pialli, G:P., Broggi, M. and Prosperi, M. (1991) Dati gravimetrici nella fascia del profilo: possibili interrogativi sull' assetto della crosta e del mantello, *Studi Geologici Camerti*, Vol.spec., 1991/**1**, 41-47.

Conti, P., Costantini, A., Decandia, F.A., Elter, F.M., Gattiglio, M., Lazzarotto, A., Meccheri, M., Pandeli, E., Rau, A., Sandrelli, F., Tongiorgi, M. and Di Pisa, A. (1991a) Structural frame of the Tuscan Paleozoic: a review, *Boll. Soc. Geol. It.* **110**, 523-541.

Conti, P., Gattiglio, M. and Meccheri, M. (1991b) The overprint of the Alpine tectono-metamorphic evolution on the Hercynian orogen: an example from the Apuane Alps (Northern Apennines, Italy), *Tectonophysics* **191**, 335-346.

Costantini, A., Decandia, F.A., Lazzarotto, A. and Sandrelli, F. (1987/88) L' Unità di Monticiano-Roccastrada fra la Montagnola Senese e il Monte Leoni (Toscana meridionale), *Atti Ticinensi di Scienze della Terra* **31**, 382-420

Dal Piaz, G.V. (1993) Evolution of Austro-Alpine and upper Penninic basement in the nothwestern Alps from Variscan convergence to post-Variscan extension, in J.F. von Raumer and F. Neubauer (eds.), *Pre-Mesozoic Geology in the Alps*, Springer Verlag, 327-344.

Decandia, F.A., Lazzarotto, A., Liotta, D. and Tavarnelli, E. (1994) Superfici strutturali profonde della Toscana e dell' Umbria-Marche, *Studi Geologici Camerti*, Vol. spec. 1994/**1**, 227-233.

Del Moro, A., Puxeddu, M., Radicati Di Brozolo, F. and Villa, M. (1982) Rb-Sr and K-Ar ages at temperatures of 300°-400°C from deep wells in the Larderello geothermal field (Italy), *Contrib. Mineral. Petrol.* **81**, 340-349.

Di Pisa, A., Franceschelli, M., Leoni, L. and Meccheri, M. (1985) Regional variation of the metamorphic temperatures across the Tuscanid I unit and its implications on the Alpine metamorphism (Apuan Alps, N. Tuscany), *Neues Jahrbuch Miner. Abh.* **151**, 197-211.

Di Pisa, A., Oggiano, G. and Talarico, F. (1993) Post-continental tectono-metamorphic evolution in the axial zone of the Hercynian belt in Sardinia:an example from Asinara island, in *Late orogenic extension in mountain belts*, Int. meeting, 4-6 March 1993 Montpellier (France), BRGM n. 219 spec. issue, 216.

Elter, F.M. (1993) Segnalazione di pieghe tipo "sheath" nella formazione delle Filladi e quarziti del Torrente Mersino (Colline metallifere), *Boll. Soc. Geol. It.* **112**, 557-561.

Elter, F.M. and Meccheri, M. (1994) Deformazione finita della fase D_1 dell' orogenesi alpina in Toscana meridionale: contributo dall' analisi delle *reduction spot* nell' Unità di Monticiano-Roccastrada, *Boll. Soc. Geol. It.* **113**, 445-450.

Elter, F.M. and Pandeli, E. (1990) Alpine and Hercynian orogenic phases in the basement rocks of the Northern Apennines (Larderello geothermal field, Southern Tuscany, Italy), *Eclogae Geol. Helv.* **83**/2: 241-264.

Elter, F.M. and Pandeli, E. (1991) Structural features of the metamorphic Paleozoic-Triassic sequences in deep geothermal drillings in the Mt. Amiata area (SE Tuscany, Italy), *Boll. Soc. Geol. It.* **110**, 511-522.

Elter, F.M. and Pandeli, E. (1993) Alpine tectono-metamorphic framework of the Tuscan Paleozoic (southern Tuscany, Italy), *Annales Tectonicae* **7**, 71-84.

Elter, F.M. and Talarico, F. (1993) The low P/high T metamorphic overprint in the "Gneiss Group" (Larderello geothermal field, Italy): a thermal record of late Hercynian orogenic extension in the basement of southern Tuscany, in *Late orogenic extension in mountain belts*, Int. meeting, 4-6 March 1993 Montpellier (France), BRGM n. 219 spec. issue, 62-63.

Elter, F.M., Franceschelli, M., Ghezzo, C., Memmi, I. and Ricci, C.A. (1986) The geology of northern Sardinia, in *Guide-book to the excursion on the Paleozoic basement of Sardinia*", IGCP n. 5, *Newsletter*, 87-102.

Franceschelli, M., Leoni, L., Memmi, I. and Puxeddu, M. (1986) Regional distribution of Al-silicates and metamorphic zonation in the low-grade Verrucano metasediments from the northern Apennines, Italy, *J. Metamorph. Geol.* **4**, 309-321.

Ghezzo, C. and Orsini, J. B. (1982) Lineamenti strutturali e composizionali del batolite ercinico sardo-corso in Sardegna, in Guida alla Geologia del Paleozoico Sardo,. Guide Geologiche Regionali S:G:I, 165-181.

Giannini, E. and Lazzarotto, A. (1975) Tectonic evolution of the Northern Apennines, in C.H. Squyres (ed.), Geology of Italy, P.E.S.L., Libia, 2,237-287.

Gualtieri, L.,Valenti, L. and Cassinis, R. (1992) Modelling gravimetrico e magnetico nelle regione di transizione tra la Liguria orientale e la Toscana, Studi Geologici Camerti, Vol. spec., 1992/2, 107-114.

Kligfield, R. (1979) The Northern Apennines as a collisional orogen, Am. J. Science 279, 676-691.

Kligfield, R., Hunziker, J., Dallmeyer, R. and Schamel, S. (1986) Dating of deformation phases using the K/Ar and 40Ar/39Ar techniques: results from northern Apennines, J. Struct. Geol. 8, 781-798.

Kornprobst, J. and Vielzeuf, D. (1984) Transcurrent crustal thinning: a mechanism for the uplift of deep continental crust/upper mantle association, in J. Kornprobst (ed.), Kimberlites II: The mantle and crust-mantle relatioships, Elsevier, Amsterdam, 349-358.

Kretz, R. (1983) Symbols for rock forming minerals, *Amer. Min.* **68**, 227-279.

Mercier, L., Johan, V., Lardeaux, J.M. and Ledru, P. (1991) Evolutions tectono-metamorphiques des nappes de l' Artense (Massif Central Francais): nouveaux marqueurs de la collision dans la chaine Varisque, *Bull. Soc. Geol. France* **163** (3), 293-308.

Moratti, G., Sani, F., Poccianti, C. and Rosselli, S. (1989) Analisi strutturale della zona di Gorfigliano (Alpi Apuane Settentrionali), *Boll. Soc. Geol. It.* **108**, 225-267.

Pandeli, E., Bertini, G. and Castellucci, P. (1991) The tectonic wedges complex of the Larderello area (southern Tuscany), *Boll. Soc. Geol. It.* **110**, 621-629.

Pandeli, E., Gianelli, G., Puxeddu, M. and Elter, F. M. (1994) The Paleozoic basement of the Northern Apennines: stratigraphy, tectono-metamorphic-sedimentary evolution and Alpine hydrothermal processes, *Mem. Soc. Geol. It.* **48**, in press.

Ricci, C.A. (1992) From crustal thickening to exhumation: petrological, structural and geochronological records in the crystalline basement of Northern Sardinia, in L.Carmignani and F.P. Sassi (eds.) *Contribution to the geology of Italy with special regard to the Paleozoic basements - A volume dedicated to Tommaso Cocozza,* IGCP N.276, *Newsletter* **5**: 187-197.

Sassi, F.P. and Zanferrari (eds.) (1989) Pre-Variscan and Variscan events in the Alpine-mediterranean belt - Stratigraphic correlation forms, *Rend. Soc. Geol. It.* **12**, 96-433.

Sestini, G. (ed.) (1970) Development of the Northern Apennines geosyncline, *Sedim. Geol.* **4**, 205-649.

Vai, G.B. and Cocozza, T. (1986) Tentative schematic zonation of the Hercynian chain in Italy, *Bull. Soc. Géol. France* **8**(2), 95-14

Vai, G.B. (1978) Tentative correlation of Paleozoic rocks, Italian Peninsula and Islands, *Oest. Akad. Wiss. Schrift. Erdw. Kom.* **3**, 313-329.

REMARKS ON A DEEP CRUSTAL SECTION ACROSS THE NORTHERN APENNINES, ITALY

M. COLI[1] AND F. CASELLI[2]
[1]*Dip. Scienze Terra, via G. La Pira 4, 50129, Firenze, Italy.*
[2]*GEOVESTA Consulting, via Magliabechi 7, 50122, Firenze, Italy.*

ABSTRACT
The crustal setting of the northern Apennines is outlined by a key crustal section cutting across the Apuan Alps metamorphic core complex, Tuscan and Ligurian nappes and the Padana buried belt, prepared using available geological and geophysical data, and made consistent with an extensional tectonic model. The key section was prepared keeping in mind that the northern Apennines resulted from the interaction of compression and extensional tectonics, which migrated from west to east as out-of-phase pulses. Data interpretation pointed out some problem areas that to be addressed to match the geological and geophysical data with the extensional tectonics models. The problem areas include:

a) the geometry of the normal faults which fragment the compressive structures of the northern Apennines at depth in order to determine which extensional models relate the normal faults to the crustal extension;
b) the seismic reflectivity or transparency of deep crustal shear-zones, and the role they play in compressional and extensional tectonics;
c) the nature and meaning of the crust-mantle transition in the inner side of the mountain belt where a melting occurs;
d) what the random deep reflectors with Vp = 8 km/s in the inner side of the mountain belt, below the anomalous crustal-mantle transition represents.

In drawing the key crust section geometric constraints required us to develop a solution to each it these problem areas; therefore the key section not only represents a summary of the available data but also an interpretation of the deep crustal tectonics of the northern Apennines

1. Introduction

In order to outline the problems related to a reconstruction of the present crustal setting of the northern Apennines, using available geological and geophysical data, a key crustal section was prepared across the Apuan Alps metamorphic core complex (Coli, 1989), the Tuscan and Ligurian nappes (Abbate et al., 1970) and the Padana buried belt (Boccaletti et al., 1985) (Fig.1).

O. Oncken and C. Janssen (eds.), Basement Tectonics 11, 91–99.

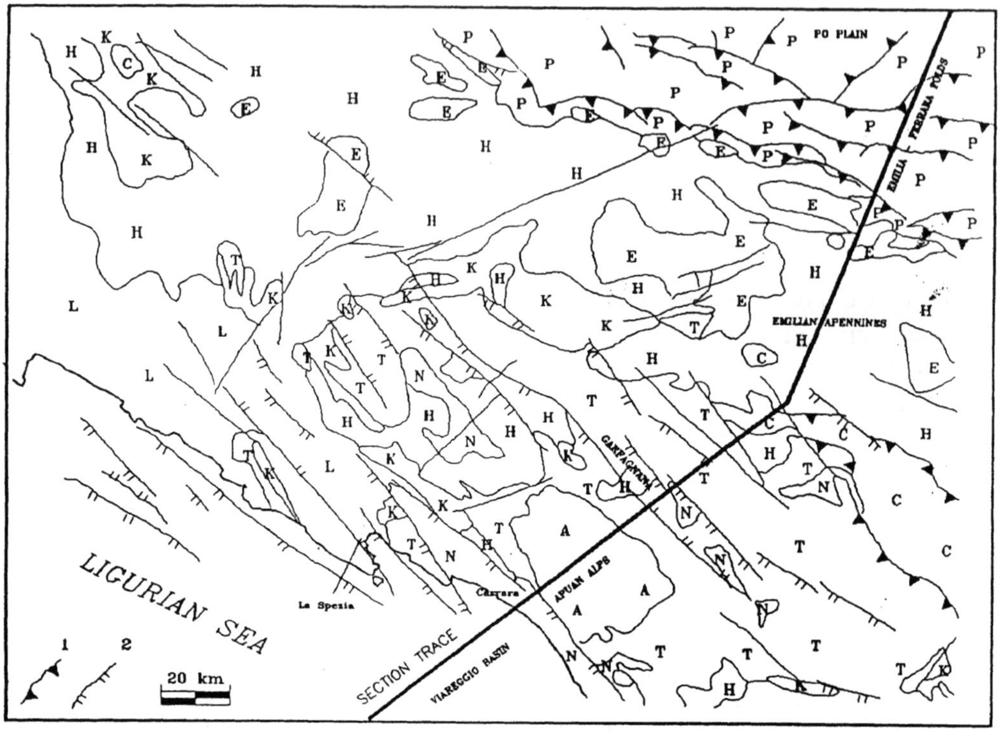

Figure 1. Tectonic sketch map of the northern Apennines. Map legend - P: Po Plain deposits; N: Neogene basin deposits; E: Epiligurian units; H: Ligurian Helminthoid flysch units; L: Ligurian ophiolitic units; K: Canetolo complex; C: Cervarola units; T: Tuscan nappe; A: Apuan Alps metamorphic unit. Symbols - 1: thrust; 2: normal fault.

1.1 GEOLOGICAL SETTING

The northern Apennines formed as a result of compressional and extensional tectonism, which migrated from west to east as out-of-phase pulses (Boccaletti et al.,1990). Compressional deformation began in the Late Cretaceous in the Piedmont- Ligurian ocean (Boccaletti et al., 1971; Channell et al., 1979). Following the formation of an oceanic suture in the Eocene, the deformation began to affect the Adria continental margin (Kligfield, 1979; Boccaletti et al., 1980), which is still underthrusting the Padana Valley to the west (Boccaletti et al., 1985).

Tyrrhenian extensional tectonism, active since Middle Miocene (Elter et al., 1975; Malinverno and Ryan, 1986), moved across the previous compressional structures with a progressive eastward shifting. Its active front currently corresponds to the main water-divide of the northern Apennines chain (Boccaletti et al., 1985; Coli, 1990). At present, the Tyrrhenian side of the belt is being affected by extensional tectonism, while the Padano-Adriatic side is being affected by compressional tectonism (Bartolini et al., 1983; Boccaletti et al., 1985).

2. The key crustal section

The key crustal section (Fig.2) across the northern Apennines extends from Cape Corse High, north of the Cape Corse promontory, through several basins and tectonic units: including the post-Tortonian (Morelli et al., 1977) Viareggio basin in the Ligurian Sea, the Apuan Alps metamorphic core complex -metamorphosed in the Late Oligocene-Early Miocene and exhumed in the Late Miocene (Kligfield et al., 1986; Coli, 1989; Carmignani and Kligfield, 1990)-, the Garfagnana Middle?-Late Pliocene (Federici and Rau, 1980) structural low, to the main body of the Tuscan and overlying Ligurian nappes in the Emilian Apennines (Abbate et al.., 1970), and the Late Pliocene-Quaternary Emilia and Ferrara folds (Pieri and Groppi, 1981) buried under the Quaternary deposits of the Po Valley. The geological and geophysical data used to prepare the cross-section, include published data from Abbate et al. (1970), A.G.I.P. (1981), Biella et al. (1987), Boccaletti and Coli eds. (1980; 1982); Boccaletti et al. (1985), Buness (1985), Buness and Giese (1990), Calcagnile and Panza (1979), Cassano et al. (1986), Castellarin et al. (1985), C.N.R. (1983), Coli et al. (1991a), Egger et al. (1988), Eva et al. (1978, 1990), Fanucci and Nicolich (1984), Giese (1985; 1990), Giese et al. (1978; 1981), Ginzburg et al. (1986), Letz et al. (1978), Morelli et al. (1977), Nadir (1988), Nicolich (1985), Panza (1985), Pieri and Groppi (1981), Roeder (1990), Scarascia (1989), Scarascia and Pellis (1985).

In the construction of the cross-section theoretical considerations for the modelling and the geometry of thrust belts and extensional basins have been taken into account (Brun and Choukroune, 1983; Boyer and Elliot, 1982; Butler, 1986; Chapple, 1978; Elliot, 1976; Gibbs, 1984; 1987; Jackson, 1987; Lister et al., 1986; Lister and Davis, 1989; Wernicke and Burchfiel, 1982; Reston, 1988; Roeder, 1980). Also the suggestions by Sclater and Christie (1980) and Passchier (1986), on seismic detectability of steep- and low-angle-normal faults and mylonites, and of Sibson (1985), on fault reactivation, have been considered.

Normal faults are supposed to be listric according to Nicolich (1987) and Bernini et al. (1991), and basement-cover relationships are set according to the field data of Ramsay et al. (1983) and Coward (1983).

3. Geometric constrains

The interpretation of geological and geophysical data in the section takes into account the following factors, which constrain the extensional tectonic model for the northern Apennines:

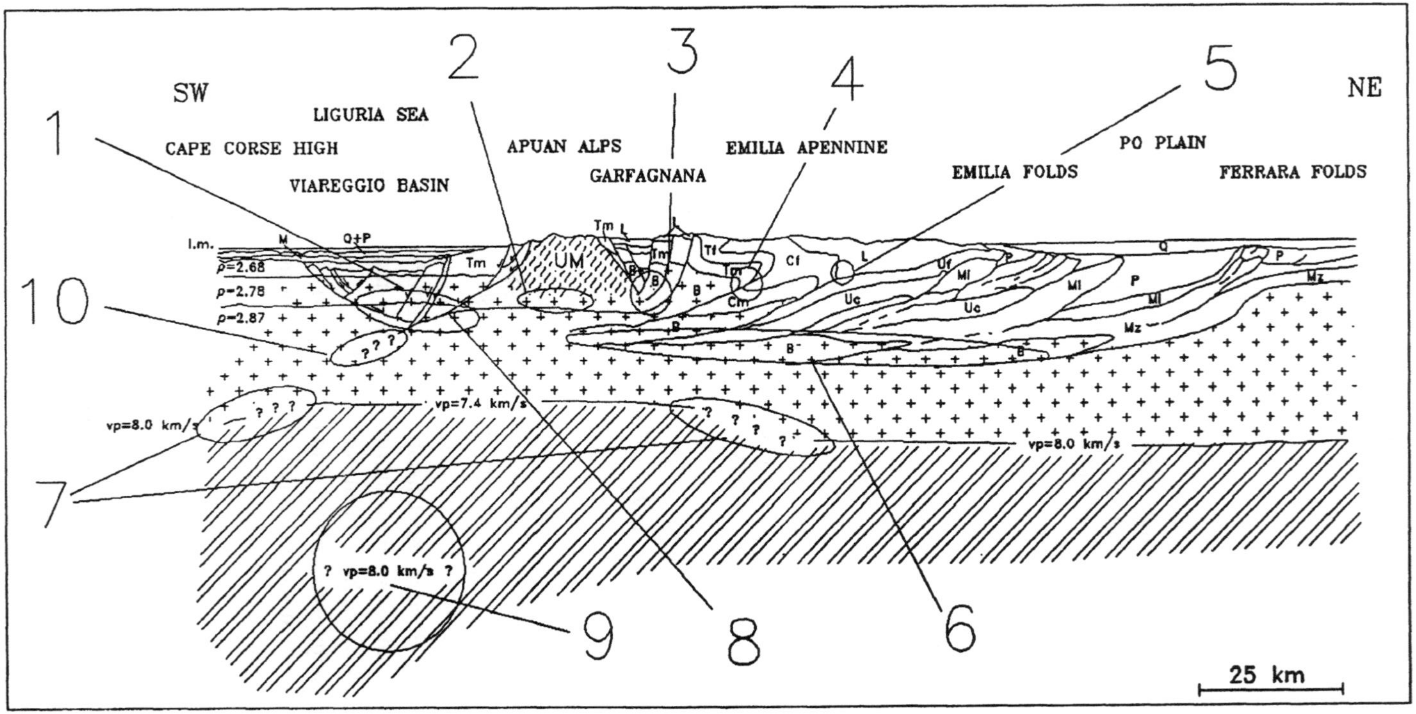

Figure 2. Key crustal section across the northern Apennines. Tectonic units: Q= Quaternary; P= Pliocene; M= Miocene; L= Ligurian units; UM= Apuan Alps metamorphic units; Tm= Tuscan nappe, carbonates; Tf= Tuscan nappe, arenaceous turbidites; Cm= Cervarola units, carbonates; Cf= Cervarola units, arenaceous turbidites; Uc= Umbrian units, carbonates; Uf= Umbrian units, arenaceous turbidites; Mi= Miocene external Clastic units; Mz= Mesozoic carbonate units; B= Basement. For numbers refer to the discussion in §4 of the text.

a) the relation of the normal faults which fragment the compressive belt, to deep crustal extension;
b) the role played in compressional and extensional regimes by deep crustal shear-zones, and the seismic reflectivity or transparency of these zones;
c) the nature and significance of the crust-mantle transition in the inner side of the mountain belt, where melting has been inferred;
d) what the random deep reflectors with Vp= 8 km/s in the inner side of the mountain belt below the anomalous crust-mantle transition represent.

4. Remarks

In recent years several efforts have been made to develop a model of extension tectonics for the northern Apennines (Lavecchia, 1988; Locardi, 1988; Carmignani and Kligfield, 1990; Coli et al., 1991a).

In the crustal section discussed here, a series of issues about the deep crustal setting of the northern Apennines, have been addressed by enhancing single geometrical factors as much as possible in order to correlate the data by a predetermined model as those mentioned above. These issues, listed below, are addressed in the cross section areas numbered 1-10 in Figure 2:

1) the inferred listric geometry of the normal faults, in the larger extensional basins is very difficult to reconcile with seismic data;
2) the tectonic unit which lies under the Apuan Alps metamorphic core complex and the crystalline basement involvement into the ductile nappes of the Apuan Alps unit;
3) the geometric relationship between the normal faults and the supposed sole-fault at the brittle-ductile transition zone;
4) whether the Cervarola unit extends below the Tuscan nappe or represents only a thin flysch unit of unknown paleogeographic origin;
5) how the Ligurian units are involved in the deep thrust of the foreland fold and thrust belt;
6) the involvement of the magnetic basement in the deep crustal thrust of the buried belt, and the possibility that the thrust was reactivated as an extensional sole-fault;
7) compressional area and the extensional area as related to crustal layering interpretation of the difference in the seismic response between the crust of the Mengel and Kern (1990);
8) the type of upper mantle present below the extensive area, where seismic data suggest a zone of melting between crust and mantle material. This is the source of the Apenninic magmas that have an unequivocal sub-crustal origin but displaying crustal- like composition signatures, and crustal melts (Peccerillo, 1990);

9) what the random seismic reflectors (Vp = 8 km/s) recorded at depth below the extensional area represent;

10) extensional tectonic models of the lower crust may be a pure-shear, domino, delamination or lozenges type, or a mixing combination of these models as hypothesised by Coli et al. (1991a)?

5. Conclusion

Analysing and interpreting the northern Apennines as a whole geodynamic system, has provided answers to a number of questions about the mountain belt.

Previous studies addressed only single and local topics detached from the general geologic framework and, therefore, failed to answer the question. It is clear that the integration of all geological, including magmatic, of Coli at al. (1991b), data and geophysical data and the use of geometric constrains in understanding the relation between surface and deep crustal structures are critical in the geodynamic modelling of the northern Apennines orogenic belt.

Acknowledgement

This research was supported by C.N.R. (National Research Council of Italy), Centro di studio per la geologia dell'Appennino in rapporto alle geosinclinali mediterranee; pubbl. n°266

References

Abbate E., Bortolotti V., Maxwell J.C., Merla G., Passerini P. and Sagri M. (1970) Development of the northern Apennines geosyncline. *Sedimentary Geology*, special issue 4, 201-648.

A.G.I.P. s.p.a. (1981) *Carta magnetica d'Italia a scala 1:500.000*. A.G.I.P. s.p.a., Milano.

Bartolini C., Bernini M., Carloni G.C., Costantini A., Federici P.R., Gasperi G., Lazzarotto A., Marchetti G., Mazzanti R., Papani G., Pranzini G., Rau A., Sandrelli F., Vercesi P.L., Castaldini D., Francavilla F. (1983) Carta neotettonica dell'Appennino Settentrionale: Note illustrative. *Boll. Soc. Geol. It.*, **101**, 523-549.

Bernini M., Papani G., Dall'Asta M., Lasagna S. and Heida P. (1991) The upper Magra Valley extension basin: a cross section between Orsaro Mountain and Zeri (Massa province). *Boll. Soc. Geol. It.*, **110**, 451-458.

Biella G.C., Gelati R., Maistrello M., Mancuso M., Massiotta P., Scarascia S. (1987) The structure of the Upper crust in the Alps- Apennines boundary region deduced from refraction seismic data. *Tectonophysics*, **142**, 71-85.

Boccaletti M., Calamita F., Deiana G.,Gelati R., Massari F., Moratti G., Ricci Lucchi F. (1990) Migrating foredeep-thrust belt system in the northern Apennines and southern Alps. *Paleo. Paleo. Paleo.*, **77**, 3-14.

Boccaletti M. and Coli M. (eds.) (1980) Sezioni geologico-strutturali in scala 1:200.000 attraverso l'Appennino Settentrionale. *C.N.R., Prog. Fin. Geod.*, pubbl.320.

Boccaletti M. and Coli M. (eds.) (1982) Carta strutturale dell'Appennino Settentrionale in scala 1:250.000. *C.N.R., Prog. Fin. Geod.*, pubbl.429.

Boccaletti M., Coli M., Decandia F.A., Giannini E., Lazzarotto A. (1980) Evoluzione dell'Appennino Settentrionale secondo un nuovo modello strutturale. *Mem. Soc. Geol. It.*, **21**, 359-373.

Boccaletti M., Coli M., Eva G., Ferrari G., Giglia G., Lazzarotto A., Merlanti F., Nicolich R., Papani G. and Postpischl D. (1985) Considerations on the seismotectonics of the northern Apennines. *Tectonophysics*, **117**, 7-38.

Boccaletti M., Elter P. and Guazzone G. (1971) Plate tectonic model for the development of the western Alps and northern Apennines. *Nature Phys. Sci*, **234**, 108-111.

Boyer S.E. and Elliot D. (1982) Thrust systems. *Bull. Am. Ass. Petrol. Geol.*, **66**, 1196-1230.

Brun J.P. and Choukroune P. (1983) Normal faulting, block tilting, and decollement in a stretched crust. *Tectonics*, **2/4**,345-356.

Buness H. (1985) The transition between the Corso-Sardinian block and the Adriatic plate. *II E.G.T. Workshop, Venice 7-9/2/85, E.S.F.*, Galson D.A. and Mueller St. Eds., Proceedings, 215-220.

Buness H. and Giese P. (1990) Acrustal section through the northwestern Adriatic plate. In:Freeman R., Giese P. and Mueller St. Eds. *The European Geotraverse: Integrative studies*. Europ. Scie. Fund., 297-304.

Butler R.W.H. (1986) Thrust tectonics, deep structures and crustal subduction in the Alps and Himalayas. *J. Geol. Soc. London*, **143**, 857-873.

Calcagnile G. and Panza G.F. (1979) Crustal and upper mantle structures beneath the Apennines region as inferred from the study of Rayleigh waves. *J. Geophysics*, **45**, 319-327.

Carmignani L. and Kligfield R. (1990) Crustal extension in the northern Apennines: the transition from compression to extension in the Alpi Apuane core complex. *Tectonics*, **9**, 1275-1303.

Cassano E., Anelli L., Fichera R. and Cappella V. (1986) *Pianura Padana: interpretazione integrata di dati geofisici e geologici*. A.G.I.P. s.p.a., Milano, 27 pp.

Castellarin A., Eva C., Giglia G. and Vai G.B. (1985) Analisi strutturale del Fronte Appenninico Padano. *Gior. Geol.*, sez. 3, **47**, 1/2, 47-76.

Chapple W.M. (1978) Mechanics of thin-skinned fold- and thust belts. *Geol. Soc. Am. Bull.*, **89**, 1189-1198.

Channell J.E., D'Argenio B. and Horvath F. (1979) Adria, the African promontory, in Mesozoic Mediterranean paleogeography. *Earth. Sci. Rev.*, **15**, 213-292.

C.N.R. (1983) Neotectonic map of Italy. *Prog. Fin. Geod.*, pubbl.114, vol.4, Quad. Ric. Scie..

Coli M. (1989) Times and mode of uplift of the Apuane Alps metamorphic complex. *Atti Ticin. Sc. Terra*, **1**, 47-56.

Coli M. (1990) Extension tectonics in the northern Apennines: a working hypothesis. *Acta Naturalia*, **26**, 81-91.

Coli M., Nicolich R., Principi G. and Treves B. (1991a) Crustal delamination of the northern Apennines thrust belt. *Boll. Soc. Geol. It.*, **110**, 501-510.

Coli M., Peccerillo A. and Principi G. (1991b) Evoluzione geodinamica recente dell'Appennino Settentrionale e attività magmatica Tosco-Laziale: vincoli e problemi. *Studi Geol. Camerti*, **Vol. Spec. 1991/1**, 403-412.

Coward M.P. (1983) Thrust tectonics, thin skinned or thick skinned, and the continuation of thrust to deep in the crust. *J. Struct. Geol.*, **5**, 2, 113-123.

Egger A., Demartin M., Ansorge J., Banda E. and Maistrello M. (1988) The gross structure of the crust under Corsica and Sardinia. *Tectonophysics*, **150**, 363-389.

Elliot D. (1976) The motion of thrust sheets. *J. Geph. Res.*, **81**, 5, 949-963.

Elter P., Giglia G., Tongiorgi M., Trevisan L. (1975) Tensional and compressional areas in the recent (tortonian to present) evolution of the northern Apennines. *Boll. Geof. Teor. Appl.*, **17**, 3-18.

Eva C., Giglia G., Graziano F. and Merlanti F. (1978) Seismicity and its relation with surface structures in the north-western Apennines. *Boll. Geof. Teor. Appl.*, **20**, 263-277.

Eva C., Augliera p., Cattaneo M. and Giglia G. (1990) Some considerations on seismotectonics of northwestern Italy. In: Freeman R., Giese P. and Mueller St. Eds. *The European Geotraverse: Integrative studies*. Europ. Scie. Fund., 289-296.

Fanucci F. and Nicolich R. (1984) Il Mar Ligure: nuove acquisizioni sulla natura, genesi ed evoluzione di un "bacino marginale". *Mem. Soc. Geol. It.*, **27**, 97-110.

Federici P.R. and Rau A. (1980) Note illustrative della neotettonica del Foglio 96 - Massa. *C.N.R., P.F. Geodinamica*, pubbl. n.356, 1365-1382.

Gibbs A.D. (1984) Structural evolution of extensional basin margins. *J. Geol. Soc. London*, **141**, 609-620.

Gibbs A.D. (1987) Development of extension and mixed-mode sedimentary basin. *J. Geol. Soc. London*, **special pubbl. n.28,** 19- 33.
Giese P. (1985) The structure of the upper lithosphere between the Ligurian Sea and the southern Alps. Part B - The consolidated crust and the uppermost mantle. In: D.A. Galson and Mueller St. Eds. - *Proceedings of the Workshop on the EGT, Venice,* 143-153.
Giese P. (1990) Recent crustal structure and young tectonics along the EGT-Segment southern Alps-Liguria Sea: an attempt of a balancing study. In: Freeman R., Giese P. and Mueller St. Eds. *The European Geotraverse: Integrative studies*. Europ. Scie. Fund., 305-310.
Giese P., Morelli C., Nicolich R. (1978) *Review of the crustal structure of the northern Apennines, The Ligurian Sea, and Corsica.* In: Cloos H. et alii Eds., Alps, Apennines, Hellenides; Inter- Union Comm. Geod., Sc., Stuttgart, Rep. 38, 221-225.
Giese P., Wigger P., Morelli C. and Nicolich R. (1981) *Seismische studien zur bestimmung der krustenstructur im bereic der geothermischen anomalie der Toskana.* Comm. Eur. Com., EUR 75-78 de MF, 108 pp.
Ginzburg A., Makris J. and Nicolich R. (1986) - European Geotraverse: a seismic refraction profile across the Ligurian arcs. *Tectonophysics*, **126**, 85-97.
Jackson J.A. (1987) Active normal faulting and crustal extension. *J. Geol. Soc. London*, **special pubbl. n.28**, 3-17.
Kligfield R. (1979) The northern Apennines as a collision orogen. *American Jour. of Sci.*, **279**, 676-691.
Kligfield R., Hunziker J., Dallmeyer R.D. and Shamel s. (1986) Dating of deformation phases using K-Ar and $^{40}Ar/^{39}Ar$ techniques: results from the northern Apennines. *J. Struct. Geol.*, **8**(7), 781-798.
Lavecchia G. (1988) The Tyrrhenian-Apennines system: structural setting and seismotectogenesis. *Tectonophysics*, **147**, 263-296.
Letz H., Reichert C., Wigger P., Giese P. (1978) *Seismic refraction measurements in the Ligurian Sea and in the northern Apennines.* In: Cloos H. et alii Eds., Alps, Apennines, Hellenides; Inter- Union Comm. Geod., Sc., Stuttgart, Rep. 38, 215-220.
Lister G.S. and Davis G.A. (1989) The origin of metamorphic core complexes and detachement faults formed during Tertiary continental extention in the northern Colorado River region, USA. *Jour. Struct. Geol.,* **11**, 1/2, 65-94.
Lister G., Etheridge M.A. and Symonds P.A. (1986) Detachment faulting and the evolution of passive continental margin. *Geology,* **14**, 246-250.
Locardi E. (1988) The origin of the Apenninic arcs. *Tectonophysics*, **146**, 105-123.
Malinverno A. and Ryan B.F. (1986) Extension in the Tyrrhenian Sea and shortening in the Apennines as result of arc migration driven by sinking of the lithosphere. *Tectonics*, **5**, 227-245.
Mengel K. and Kern H. (1990) Petrologic versus seismic Moho and crustal root mysteries of the Hercynian orogen. Freeman R., Giese P. and Mueller St. Eds. *The European Geotraverse: Integrative studies*. Europ. Scie. Fund., 169-176.
Morelli C., Giese P., Carrozzo M.T., Colombi B., Guerra I., Hirn A., Letz H., Nicolich R., Prodehl C., Reichert C., Roewer P., Sapin M., Scarascia S. and Wigger P. (1977) Crustal and upper mantle structure of the northern Apennines, the Ligurian Sea and Corsica, derived from seismic and gravimetric data. *Boll. Geof. Teor. Appl.*, **19**, 199-260.
Nadir S. (1988) *Structure de la croute continentale entre les Alpes occidentale at les Alpes ligures.* These, Univ. Paris, VI, 120 pp.
Nicolich R. (1985) EGT southern segment: reflection seismics in the offshore areas. In: Galson D.A. and Mueller St. Eds., *II E.G.T. Workshop, Venice 7-9/2/85, E.S.F.*, Proceedings, 33-38.
Nicolich R. (1987) *Crustal structures from seismic studies in the frame of the EGT, southern segment, and CROP projects.* Acc. Naz. Lincei, pre-print vol. of "The Lithosphere in Italy: advances in Earth Sciences research", Rome 5-6/5/87, 1-14.
Panza G.F. (1985) Lateral variations in the Lithosphere in correspondance of the southern segment of the EGT. *II E.G.T. Workshop, Venice 7-9/2/85, E.S.F.,* Galson D.A. and Mueller St. Eds., Proceedings, 47-52.
Passchier C.W. (1986) Mylonites in the continental crust and their role as seismic reflectors. *Geol. en Mijnbouw*, **65**, 167-176.
Pieri M. and Groppi G. (1981) Subsurface geological structure of the Po plain, Italy. *Prog. Fin. Geod.*, pubbl.414, 13 pp.

Peccerillo A. (1990) On the origin of the Italian potassic magmas - Comments. *Chemical Geology*, **85**, 183-196.

Ramsay J.G., Casey M. and Kligfield R. (1983) - Role of shear in development of the Helvetic fold-thrust belt of Switzerland. *Geology*, **11**, 8, 439-442.

Reston T.J. (1988) Evidence for shear zones in the lower crust offshore Britain. *Tectonics*, **7**, 929-945.

Roeder D. (1980) Geodynamics of the Alpine-Mediterranean system, a sysnthesis. *Eclogae Geol. Helv.*, **73**, 2, 353-377.

Roeder D. (1990) Crustal structure and kinematics of the Ligurian and west-alpine regions. In: Freeman R., Giese P. and Mueller St. Eds. *The European geotraverse: Integrative studies*. Europ. Scie. Fund., 311-326.

Scarascia S. (1989) *Crustal structures under the northern Apennines and central Alps from additional data obtained in the E.G.T. '86 seismic experiments (a report)*. E.U.G., Strasburgo.

Scarascia S. and Pellis G. (1985) Crustal structure of the northern Apennines. Part A: The upper crust. In: Galson D.A. and Mueller St. Eds. - *Proceedings of the Workshop on the EGT, Venice, ESF*, 137-142.

Sclater J.G. and Christie P.A. (1980) Continental stretching and explanation of the post mid-Cratceous subsidence of the central North Sea basin. *J. Geoph. Res.*, **85**, 3711-3739.

Sibson R.H. (1985) A note on fault reactivation. *J. Struct: Geol.*, **7**, 6, 751-754.

Wernicke B. and Burchfiel B.C. (1982) Modes of extensional tectonics. *J. Struct. Geol.*, **4**, 2, 105-115.

MULTIFRACTAL ANALYSIS OF FAULT STRUCTURES IN BASEMENT ROCKS

I.R.STAKHOVSKY
Institute of Physics of the Earth, Russian Academy of Sciences,
B.Gruzinskaya 10, Moscow, 123810, RUSSIA

The theory of multifractal measures is applied for analysis of the geometry of fault system of the Caucasus. By the help of computer simulation it is shown that multifractality of faulting arises due to stochastic nature of rock fracture. The statement is suggested that the scaling of faults is related to Earth's crust ability to accumulate elastic energy during tectonic deformation.

1. Introduction

Theory of fractal sets is broadly used in structural geology for investigating such self-similar processes as faulting (Hirata, 1989) and cracking in rocks (Hirata et al, 1987; Sammis and Biegel, 1989) because it provides the simplest nontrivial description of scale invariance of physical systems. Computer simulation of quasi-static fracture in solids that incorporates lattice models (Louis and Guinea, 1989; Hassold and Srolovitz, 1989) also generates fractal patterns. In geophysical applications, however, multifractal measures are more relevant than fractal sets because geophysical quantities are best described as measures (or fields). This paper presents a multifractal method for describing crack and fault patterns in rocks using computer images of crack (fault) clusters and fault systems as depicted on geologic maps.

2. Computer Simulation

The most vivid way of illustrating the multifractality of faults is a computer simulation of rock fracture. Following the methods of previous researchers (Louis and Guinea, 1989; Hassold and Srolovitz, 1989) a rectangular lattice with bonds that simulate the interaction forces among inhomogeneities was used in our work

O. Oncken and C. Janssen (eds.), Basement Tectonics 11, 101–110.

as a computational model of structured solid (rock). The deformational properties of the lattice satisfy the equations of the mechanics of solids. Initial defects were created in the lattice by "rupturing" lattice bonds. The resulting cavities imitate initial cracks in rock and become sites where stress concentrates.

Let x_1, x_2, x_3 be a system of orthogonal coordinates with x_1 and x_2 located in the plane of the lattice. The equations of stress equilibrium in solids are (using index denotements and Einstein's summation convention):

$$\sigma_{ij,j} = 0, \tag{1}$$

where σ_{ij} is the stress tensor. Eqs.(1) take into account neither the dynamic nor temperature effects during fracturing.

Later on we imply that crack growth can be defined by stress deviator behaviour. In framework of plane problem we'll consider the case when solid undergoes the state of antiplane shear. In antiplane shear by determination:

$$\sigma_{ij} = \begin{Vmatrix} 0 & 0 & \sigma_{13} \\ \cdot & 0 & \sigma_{23} \\ \cdot & \cdot & 0 \end{Vmatrix}. \tag{2}$$

If $\vec{U}$ - the displacement vector and U_1, U_2, U_3 - the displacement vector components, γ_{ij} - the strain tensor, then the shear strains are equal:

$$\gamma_{13} = U_{3,1}, \qquad \gamma_{23} = U_{3,2}. \tag{3}$$

The inferred rheology of basement rocks under tectonic deformation is firmo-viscous rheology of Kelvin body (Scheidegger, 1957). The connection between components of stress tensor σ_{ij} and strain tensor γ_{ij} corresponding to a linear firmo-viscous body (Kelvin body) can be represented for antiplane shear as:

$$\sigma_{ij} = 2 \cdot \mu \cdot \gamma_{ij} + 2 \cdot \eta \cdot d\gamma_{ij} / dt = L(2 \cdot \gamma_{ij}), \tag{4}$$

where t - time, μ - shear modulus, η - coefficient of shear viscosity, $L = \mu + \eta \cdot d/dt$ - linear visco-elastic operator. Substituting Eqs.(2),(3),(4) in system (1) one can obtain:

$$L(U_{3,11}+U_{3,22})=0 \quad \longrightarrow \quad L(\Delta U_3)=0, \tag{5}$$

where Δ is the 2-D Laplace operator. If the stress field is stationary, i.e.

$$\partial\sigma_{ij}/\partial t=0, \tag{6}$$

equation (5) can be satisfied by the expression:

$$U_3=U_3^{\infty}+(\overset{\circ}{U}_3-U_3^{\infty})\cdot exp(-t\cdot\mu/\eta), \tag{7}$$

where U_3 is the displacement directed along the normal to the lattice plane. Crack growth in lattice models is simulated by lattice bond rupture so the fracturing process occurs as descrete "steps". A "step" is interpreted to be the transition from the state with K broken lattice bonds to the state with K+1 broken lattice bonds, K=1,2,3.... In formula (7) $\overset{\circ}{U}_3$ is the displacement field at the time of "step" initiation ($\overset{\circ}{t}$) and U_3^{∞} is the displacement field at time t^{∞} when the lattice field reaches an equilibrium state. It can be shown that Eq.(7) is the solution of Eq.(5) if U_3^{∞} is a harmonic function satisfying the boundary conditions, i.e.:

$$\Delta U_3^{\infty}=0. \tag{8}$$

For numerical integration Eq.(8) was substituted with a series of finite-difference equations which were solved iteratively. Boundary conditions on the external contour of the lattice G were set as $U_3(G)=f(G)$. During the whole process of crack cluster growth the initial displacement along the external contour of the lattice remained constant producing stress on lattice bonds. On internal contour of the lattice boundary conditions were set as $\sigma_{ij}n_j=0$, where $\vec{\boldsymbol{n}}$ is the unit vector directed along the external normal to surface. Rupture of the lattice bonds led to abrupt changes in the stress field, with the Eq.(6) condition occurring between rupture events. Rupture of the lattice bonds in this model is considered to be a probabilistic fluctuational process. The fluctuational character of crack (fault) growth is inherited from the micro-level where fracturing is closely related to quantum statistics (Gilman and Tong, 1971) or thermofluctuative statistics (Zhurkov and Tomashevsky, 1966). The fracture "step" can be determined if we know the location of the bond to be broken and the time t^f that has elapsed since the previous bond rupture. With $W1$ as the quantity:

$$W1 = (t^{\infty} - t^{i})/t^{\infty}, \quad t^{\circ} \le t^{i} \le t^{f} \le t^{\infty}, \tag{9}$$

the time equal to infinity, t^{∞}, was defined as:

$$t^{\infty} = 5 \cdot \eta/\mu = 5 \cdot t^{rel}, \tag{10}$$

where t^{rel} is the time of relaxation. In accordance with the fluctuational fracture conception the value of t^f was determined with a random number generator from the interval $t^i \le t^f \le t^{\infty}$. The numerical value of $W1$ (with t^i as the lower limit of allowed interval of fracture times) was established as a model parameter before cluster generation. Under condition $W1 = 1$ any moment of time between t° and t^{∞} has an equally chance for fracturing. Condition $W1 = 0$ practically turns the model into purely elastic one.

After establishing the fracture time t^f and knowing the function U_3^{∞} from the numerical solution of Eq.(8) the state of the displacement field at the moment of bond rupture was determined with Eq.(7). Thus the lattice bond strains on internal contour of the lattice were calculated by using the finite-difference approximations of Eqs.(3). The second parameter of the model, $W2$, defines the group of the most deformed bonds:

$$W2 = \textit{Most strained bonds} \,/\, \textit{All bonds of internal contour}\,. \tag{11}$$

Parameter $W2$ can accept values of ≤ 1 but to stay consistent with physical experiment (Hirata et al, 1987; Sammis and Biegel, 1989) values $W2 \le 0.2$ are optimal. Lattice bonds were sorted based on the amount of strain. Fracture of any bond occurred after the sorting in the group of most strained bonds had equal chance. The random number generator chose one lattice bond from this group which was fractured in a current "step" of cluster generation. That changed the geometry of the internal cavity. The cluster growth process reflected repetitions of above procedure under various values of $W1$ and $W2$. The displacement field at the moment of bond rupture formed the initial conditions for next "step" of cluster generation. An example of a fracture pattern generated with this algorithm is shown in *Figure 1.*

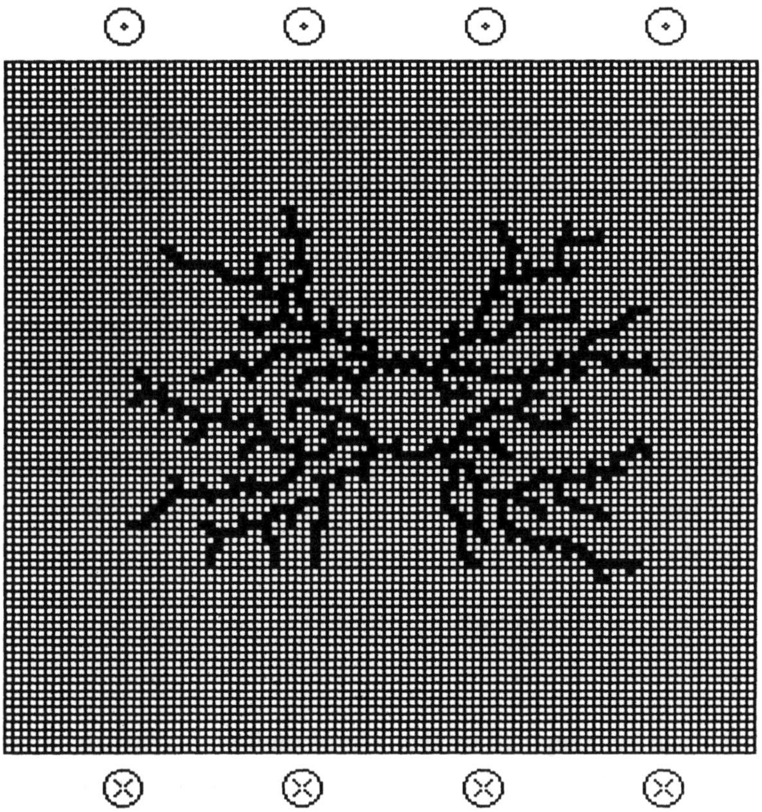

Figure 1. Crack cluster generated on the lattice 90x90 with *W1=1* and *W2=0.17*. This cluster was generated from two initial defects. It is easy to see the closed loops of cracks, the confluence of cracks and the multiple branching of cracks. Lattice cells with broken bonds are blacked in and show the trajectories of the cracks.

3. Multifractal Data Analysis Technique

In the theory of multifractal measures (Stanley and Meakin, 1988; Mandelbrot, 1989; Lovejoy and Schertzer, 1990) a set is regarded as the union of fractal subsets. The probabilistic measure of such set can be represented as the superposition of addenda - that is to say singularities of order a each supported by a fractal set with fractal dimension $f(a)$. The density of set elements can be interpreted as the realisation of multifractal probability measure in the limit of the number of set elements tending to infinity. Following (Mandelbrot, 1989), let's cover a 2-D set by a grid of square boxes. The dimensionless grid box length r can be changed through grid renormalization.

Allow

$$p_i = A_i / A, \tag{12}$$

where A_i - number of set elements in i-th box, A - all of set elements, p_i - an estimation of probability that the set elements fall into i-th box, $\sum_i p_i = 1$. The set can be characterized by the measure $M(q,r)$:

$$M(q,r) = \sum_i p_i^q \cdot r^d \xrightarrow[r \to 0]{} \begin{cases} 0, & d > \tau(q) \\ \infty, & d < \tau(q) \end{cases}, \tag{13}$$

where q and d are exponents. Finite values of measure (13) connect variables q and d by functional dependence, $d = \tau(q)$. If values of measure (13) are not equal to zero or infinity then:

$$\sum_i p_i^q \propto r^{-\tau(q)}. \tag{14}$$

Function $\tau(q)$ can be computed by the help of multiple renormalization of a box grid. Values of $\tau(q)$ under definite q are given from the slopes of linear regressions $\sum_i p_i^q$ versus r in double logarithmic scales. It is also possible to characterize the scaling of multifractal fields in terms of the order of singularity a and the spectrum of singularities $f(a)$. Setting probability p_i to depend on r by power law:

$$p_i \propto r^{a_i}, \tag{15}$$

the singularity index (Lipschitz-Hölder exponent) a_i depends on the box position. The measure of the set is multifractal if the number of boxes with the common value of a, say $\sum N_a(r)$, depends on r such that:

$$\sum N_a(r) \propto \rho(a,r) \cdot r^{-f(a)}, \tag{16}$$

where $\rho(a,r)$ is a slowly varying function of r and a. Using the Legendre transformation it is possible to transit from the function $\tau(q)$ to the singularity

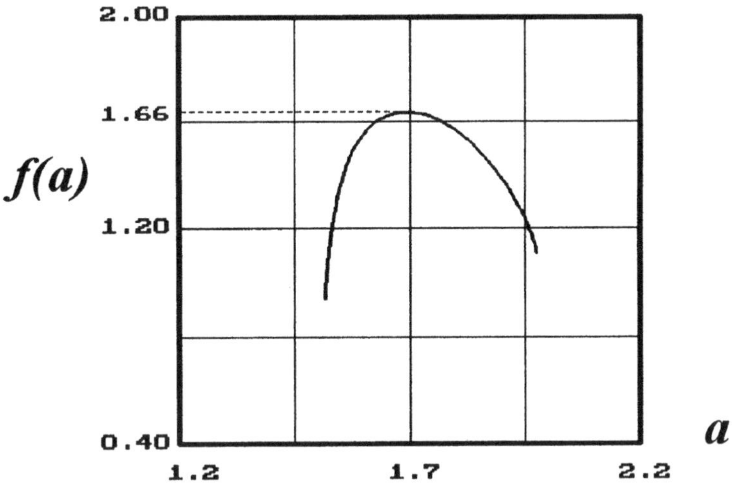

Figure 2. Singularity spectrum of fractal measure for fracture pattern from *Figure 1*. Dashed line marks the value of monofractal dimension of the set of broken lattice bonds.

spectrum $f(a)$ due to their one-to-one correspondence (Mandelbrot, 1989):

$$a(q) = -d\tau(q)/dq \quad , \qquad (17)$$
$$f(a(q)) = q \cdot a + \tau(q) .$$

Computing a singularity spectrum for fractal measure can be viewed as the aim of scaling analysis. It is worth noting that the box-counting method involves a row of smoothing operations, is sensitive to poor sampling statistics and finite grid resolution. Therefore Chhabra et al (1989) have worked out the algorithm of direct determination of $f(a)$ by renormalizing grid of boxes:

$$a(q) = \lim_{r \to 0} (1/\ln r) \cdot \sum_i \overline{p}_i(q,r) \cdot \ln p_i(r), \qquad (18)$$
$$f(a(q)) = \lim_{r \to 0} (1/\ln r) \cdot \sum_i \overline{p}_i(q,r) \cdot \ln \overline{p}_i(q,r),$$

where $$\overline{p}_i(q,r) = p_i^q(r) \Big/ \sum_j p_j^q(r)$$

Figure 2 shows the graph of scaling function for the measure of set of broken lattice bonds, which form the fracture pattern in *Figure 1,* calculated using the algorithm of direct determination of $f(a)$. Singularity spectrum characterizes the crack cluster as a structure with an infinite dimension hierarchy and the cracking process as multifractal process.

4. Clusters of Geological Faults

The largest-scale example of quasi-static fracture in rocks is faulting in the Earth's crust. *Figure 3* demonstrates the map of active faults of the Caucasus.

To perform multifractal analysis of fault field we chose three 200x200 km polygons (west, central and east polygons in *Figure 3*). We assume that every fault line is smooth enough to have the possibility to measure its length by relevant unit of measurement. By reproducing the maps on a computer screen the trajectories of faults of each polygon with the help of screen scanning were turned into series of pixels with 256x256 resolution. 2-D sets of pixels were covered by grid of boxes and the ordinary box-counting algorithm (Eqs.(12)-(17)) was applied. The whole number of grid boxes was altered from 2^0 to 2^{14}.

The graphs of functions $f(a)$ calculated for all polygons are shown in *Figure 4a*. The singularity spectrum $f(a)$ is independent of r characteristic of fractal measure. The lowest value of singularity index a corresponds to most

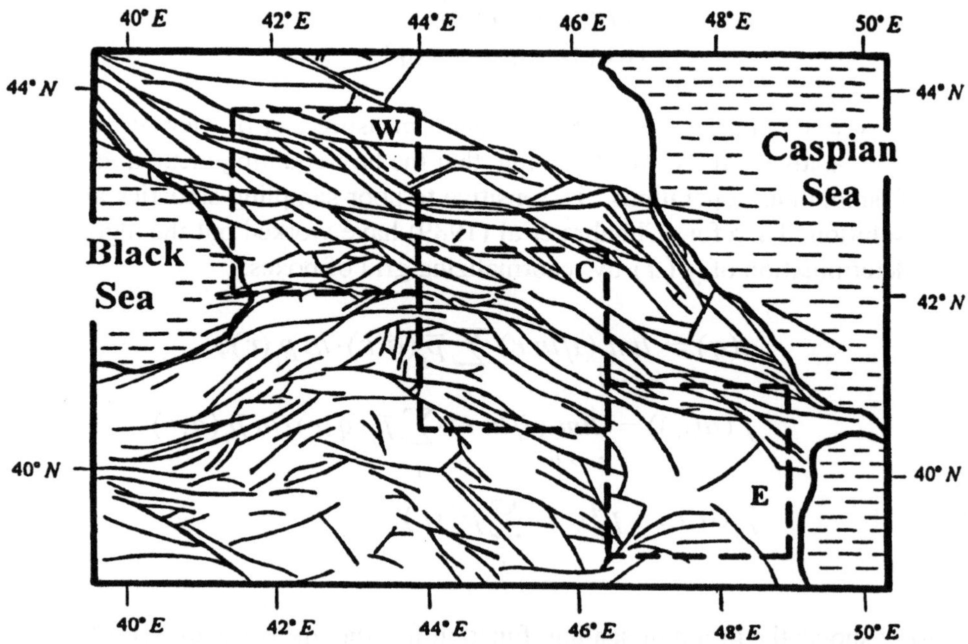

Figure 3. The map of active faults of the Caucasus. Three polygons were extracted for multifractal analysis - west (W), central (C) and east (E) - along the Main Caucasus Range. Map by T.P.Belousov (Institute of Physics of the Earth, Moscow, Russia).

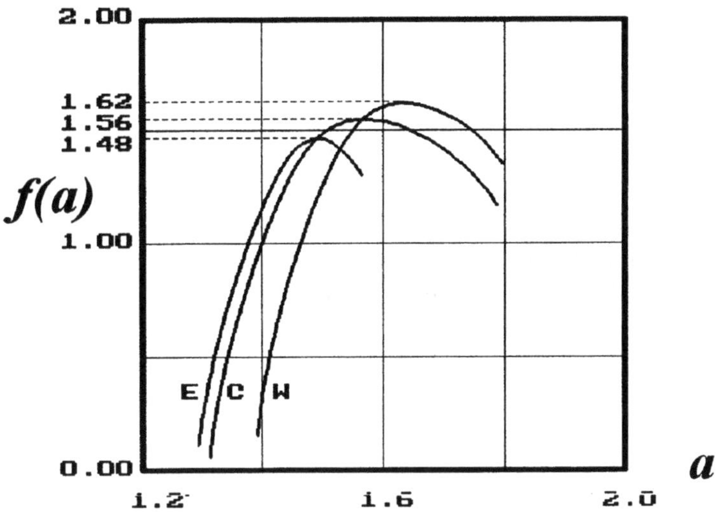

a)

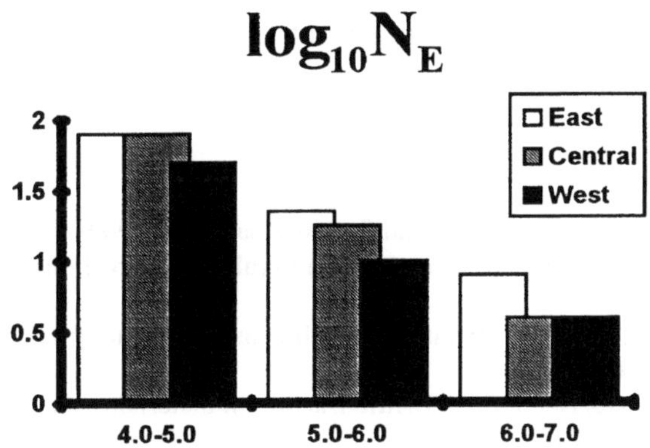

b)

Figure 4 a) Functions *f(a)* for fractal measures supported by sets of pixels discretizing the trajectories of faults: E - east polygon, C- central and W-west polygon. Singularities of measure with indices a are supported by the set with fractal dimension *f(a)*.

b) Histogram illustrating the seismic activity of all polygons. $\mathbf{N_E}$ is the number of earthquakes with various magnitude intervals.

concentrated domain while the largest value of a corresponds to most rarefied domain of fault cluster. *Figure 4a* shows that the singularity spectra do not reduce to a single point, i.e. the fault patterns haven't ideal monofractal structure.

The singularity spectra provide the "construction rule" of fault configurations and quantitatively describe the global self-similarity of the geometry of fault clusters.

It is noteworthy that fault scaling is apparently associated with seismic activity of the Earth's crust. *Figure 4b* shows seismic data for the three polygons for the period 1850-1991, taken from the catalogue of Institute of Physics of the Earth, Moscow. The polygons which differ in the scaling parameters of their fault structures also differ in seismic activity - to be exact, the higher the value of $f(a)$-spectrum extremum the lower is seismic activity of the same polygon. This indicates that the fault scaling is related to Earth's crust material ability to accumulate elastic energy during the process of tectonic deformation (and correspondingly it related to seismic statistics).

5. Conclusions

Analysis of the geometry of both model and natural fault trajectories suggests that faulting is expected to be multifractal process. The singularity spectrum $f(a)$ can be used as a quantitative scale invariant characteristic of fault system.

6. References

Chhabra A.B., Meneveau C., Jensen V., and Streenivasan K.R. (1989) Direct determination of the *f(a)* singularity spectrum and its application to fully developed turbulence. *Phys.Rev.A*, **40**, 5284-5294.

Gilman J.J. and Tong H.C. (1971) Quantum tunneling as elementary fracture process. *J.Appl.Phys.*, **42**, 3479-3486.

Hassold G.N. and Srolovitz D.J. (1989) Brittle fracture in materials with random defects. *Phys.Rev.B*, **39**, 9273-9281.

Hirata T., Satoh T., and Ito K. (1987) Fractal structures of spatial distribution of microfracturing in rocks. *Geophys.J.R.astr.Soc.*, **90**, 369-374.

Hirata T. (1989) Fractal dimension of fault systems in Japan: fractal structure in rock fracture geometry at various scales, *PAGEOPH*, **131**, 157-170.

Louis E. and Guinea F. (1989) Fracture as a growth process. *Physica D*, **38** , 235-241.

Lovejoy S. and Schertzer D. (1990) Multifractals, universality classes and satellite and radar measurements of cloud and rain fields. *J.Geophys.Res.D*, **95**, 2021-2034.

Mandelbrot B.B. (1989) Multifractal measures, especially for geophysicist. *PAGEOPH*, **131**, 5-42.

Sammis C.G. and Biegel R.L. (1989) Fractals, fault-gouge and friction. *In*: *Fractals in Geophysics*.-Basel,Boston,Berlin, 255-271.

Scheidegger A.E. (1957) Rheology of the Earth: the basic problem of geodynamics. *Canad.J.Phys.*, **35**, 4-14.

Stanley H.E. and Meakin P. (1988) Multifractal phenomena in physics and chemistry. Nature, 335, 405-409.

Zhurkov S.N. and Tomashevsky E.E. (1966) Thermal fluctuations of atoms and fracture. *Proc. conf. "Phys. Basis Yield a. Fracture"*, Oxford, 200.

LARGE-SCALE GEOMETRY OF MIGMATITES - IMPLICATIONS FOR THE SEISMIC PROPERTIES OF CONTINENTAL CRUST

K. Bierbrauer and O.Oncken
GeoForschungsZentrum Potsdam, Telegrafenberg C2,
D-14473 Potsdam

Abstract

A quantitative analysis of migmatitic textures is used to describe the complex large-scale geometry of a Variscan migmatite terrain (Argentera massif, western Alps). Small-scale textural variations in migmatites, which can be observed in hand-sample and outcrop scale, closely resemble the overall structural style and kilometer-scale distribution of texturally distinct migmatites within the Argentera basement. Three-dimensional contouring of textural properties in this area suggests a strong influence of migmatization processes on the resulting seismic image of continental crust. A three-dimensional model which has been derived from the geologically reasonable extrapolation of field data shows that regional migmatization mainly results in the formation of "textural gradients". Such gradients destroy pre-existing fabrics, lithological boundaries and layers of various acoustic impedances. Apart from the typical occurrence of granitic intrusions within the Variscan domain, the strong migmatization of basement gneisses is believed to be largely responsible for the observed seismic transparency in the mid- to upper crustal parts of the internal European Variscides.

1. Introduction and geological context

Deep seismic reflection profiles from the internal parts of the Variscan belt often display a highly reflective lower crust in contrast to a less reflective or seismically transparent middle to upper crust. It has been suggested that many lower crustal reflections may originate from fine-scale metamorphic layering, mylonite zones, and lithological layering (eg. Christensen, 1989; Hurich and Smithson, 1987). Such layering could be the result of strong ductile deformation during a late orogenic extensional event (eg. Mooney and Meissner, 1992; Rey, 1993). Other contributions to lower crustal reflectivity may be mafic intrusions (eg. Dohr and Meissner, 1975), partial melting (eg. Hale and Thompson, 1982), or the presence of fluid-filled cracks as has been suggested by Matthews and Cheadle (1986). In the corresponding experiments the relatively transparent middle and upper crustal levels, which are typical for many seismic reflection profiles from the internal Variscides, are poorly understood. Most of the present day exposed Variscan high-grade gneiss terrains presumably represent the more or less seismically transparent type of crust. These areas are generally characterized by a strong migmatization and

O. Oncken and C. Janssen (eds.), Basement Tectonics 11, 111–124.

large volumes of granites. The thermo-metamorphic evolution of these terrains must somehow be related to the observed seismic reflectivity pattern.

The Argentera massif represents a well exposed block of exhumed Variscan crust in the Helvetian domain of the western Alps. It is the southernmost exposure of a chain of several Variscan basement units, the so called "external crystalline massifs" of the Alps (see Fig. 1). The external massifs are part of the Moldanubian domain (von Raumer et al. 1993) a zone which is largely composed of high grade gneisses and migmatites from the internal deeper parts of the Variscan collisional belt. Therefore the tectono-metamorphic evolution of the external massifs is very similar to the well known Bohemian Massif, the Black Forest area or the French Massif Central region which are known as the classical Moldanubian basement exposures.

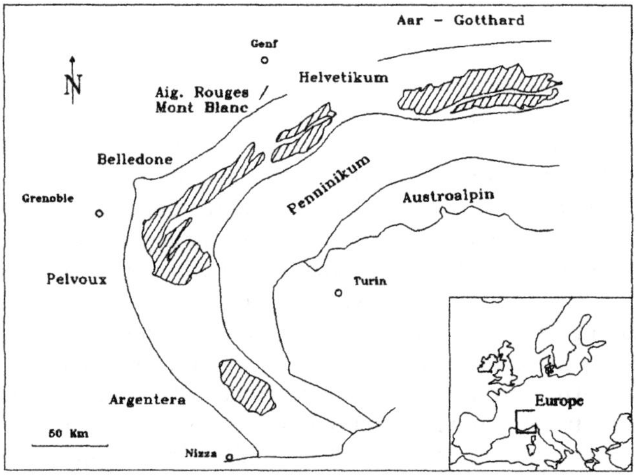

Figure 1 Distribution of Variscan crystalline basement in the external massifs of the Helvetian domain of the western Alps.

The Variscan event was primarily characterized by crustal thickening and nappe-stacking (eg. Matte, 1986; Franke, 1989) followed by a characteristic low P-/high T-metamorphism and associated regional migmatization at the thermal peak (von Raumer et al., 1993). Due to the profound thermal imprint during the main Variscan evolution the older metamorphic history of the Moldanubian domain remains poorly understood. Amphibolite facies migmatites, containing eclogites and high P-granulites have been reported from many different localities within the Moldanubian zone (eg. Bogdanoff and Latouche, 1986 in the Argentera massif; von Raumer, 1974 in the Mont-Blanc and Aiguilles-Rouges massifs; Wimmenauer and Stenger, 1989 in the Black Forest; O'Brien, 1989 and O'Brien et al., 1992 in the northwestern Bohemian Massif). However, it is not always clear whether the eclogites and related HP-granulites belong to an older pre-Variscan Cadomian basement or whether they are associated with the early Variscan crustal thickening event. Moreover different types of eclogites and granulites which have formed at different times (Pin and Vielzeuf, 1983) may be present within the Moldanubian basement. Petrological and structural data from the Argentera area suggest that large parts of the basement had been transported into a lower crustal position prior to the closure of a main Variscan suture (von Raumer, 1988; von Raumer et al., 1993). During the later event this lower crustal section was uplifted and subsequently exposed in the upper plate of the collision zone. In the Argentera basement this evolution is manifested by the development of large-scale fold structures and the disruption of an older strongly foliated metamorphic layering during the migmatization event. The large-scale geometry and the textural evolution of the resulting migmatitic gneisses are the main subjects of the following paragraphs.

2. Textural properties

A variety of textures are known from high-grade gneisses and associated migmatites. The number of descriptive terms introduced in the literature and the use and applications of this terminology by many authors make it difficult to compare geological maps of different crystalline areas, especially where no additional structural surveys are available. As a result, while the complex geometry of high-grade gneiss terrains may be familiar to the individual who studied it, there are no general structural models which adequately describe such basement areas and which can be used by other scientists.

This problem was addressed earlier by Berthelsen (1960) who developed a structural classification of gneisses (including migmatites) to map an area of 2500 square kilometers in Southwest Greenland. As many different field workers were involved during this mapping project, a classification based on structural criteria was the only possible and objective working method that could be used. Especially, after the publication of Mehnert's classical work on migmatites (Mehnert, 1968) most workers adopted Mehnert's migmatite terminology and applied his concepts of anatexis to their working areas. More recent work on migmatites was mainly focused on the processes of migmatization itself and the distinction between metamorphic versus anatectic migmatites (see Ashworth (1976) and authors therein; Yardley, 1978; McLellan, 1989).

The application of modern 3D-visualization software to study the internal architecture of migmatite terrains requires a translation of the typical appearance of gneisses and migmatites into numerical properties. Such property values can then be modelled and displayed in 3D-space. In this context the above mentioned structural concept of Berthelsen (1960) appears to be very useful, although his early classification must be modified and extended into a more dynamic and sequential evolution of textures.

The classification of textures in Fig.2 is simple but can be applied to any given type of crystalline rock. The classification basically describes the degree of layering and anisotropy in a given rock. This property decreases from type 1 to type 6. The scheme also describes the amount of textural homogenization due to migmatization phenomena or the intrusion of melt. Textural homogenization on the other hand increases from type 1 to type 6 in the classification scheme. In this sequence of textures, type 1 applies to highly foliated or layered gneisses which are typical for many high-grade metamorphic rocks, whereas type 6 is intended to characterize a homogeneous igneous rock type. Migmatites may be regarded as the intermediate stage between the two end member types, with predominantly metamorphic and igneous textures, respectively. However, this classification is descriptive and does not necessarily suggest an increasing amount of partial melt from type 1 to type 6. The igneous type 6 texture may be the result of crystallization from a melt as well as the result of metasomatic reactions or post-tectonic grain growth during the transformation of a pre-existing metamorphic fabric. A study of the existing migmatite literature shows that most migmatites developed from a pre-existing highly deformed and foliated protolith gneiss (eg. Hopgood et al. 1976 and 1978; Amit and Eyal, 1976; Yardley, 1978; Johannes and Gupta, 1982; Mehnert and Büsch 1982). As migmatization processes tends to destroy older fabrics as shown in Fig.2, this is very suitable to describe the geometrical effects of migmatization.

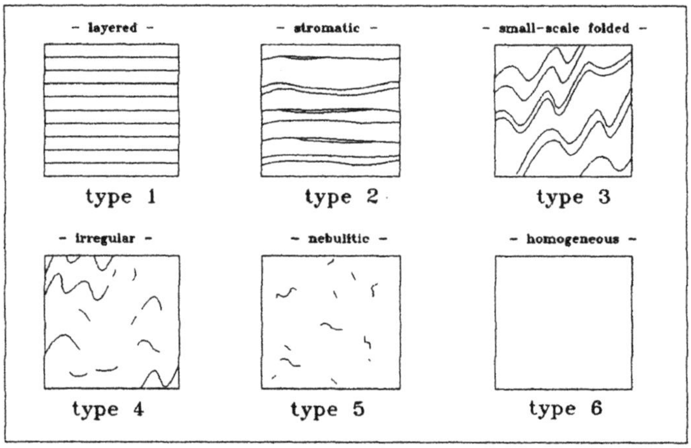

Figure 2 Classification scheme of rock textures based on geometrical properties. The diagram has been designed to describe the geometry of a strongly migmatized basement terrain. The textural classification can be used to characterize migmatites on any scale of observation. The scheme is descriptive and illustrates an decreasing degree of layering and an increasing amount of textural homogenisation from type 1 to type 6. The names for the textural classes result from the most common descriptive terms which have been used to characterize similar gneisses and migmatites in the literature.

3. Scale of observation and modelling technique

The textural mapping method strongly depends on the scale of observation. All types of textures may occur within a single outcrop. Likewise, larger outcrops may only display a very narrow range of textures. Our study was designed to visualize the textural variations which occur on the scale of the vertical resolution of reflection profiling (50 to several hundreds of meters) and to work out the structural modifications of a strong regional migmatization event. An example of how the classification scheme may be applied to a migmatite sample is illustrated in Fig.3. The size of each cell is 7 by 7 centimeters. Black numbers indicate the overall textural type within each cell. Some of the cells can be subdivided into smaller textural domains as indicated by the smaller white numbers.

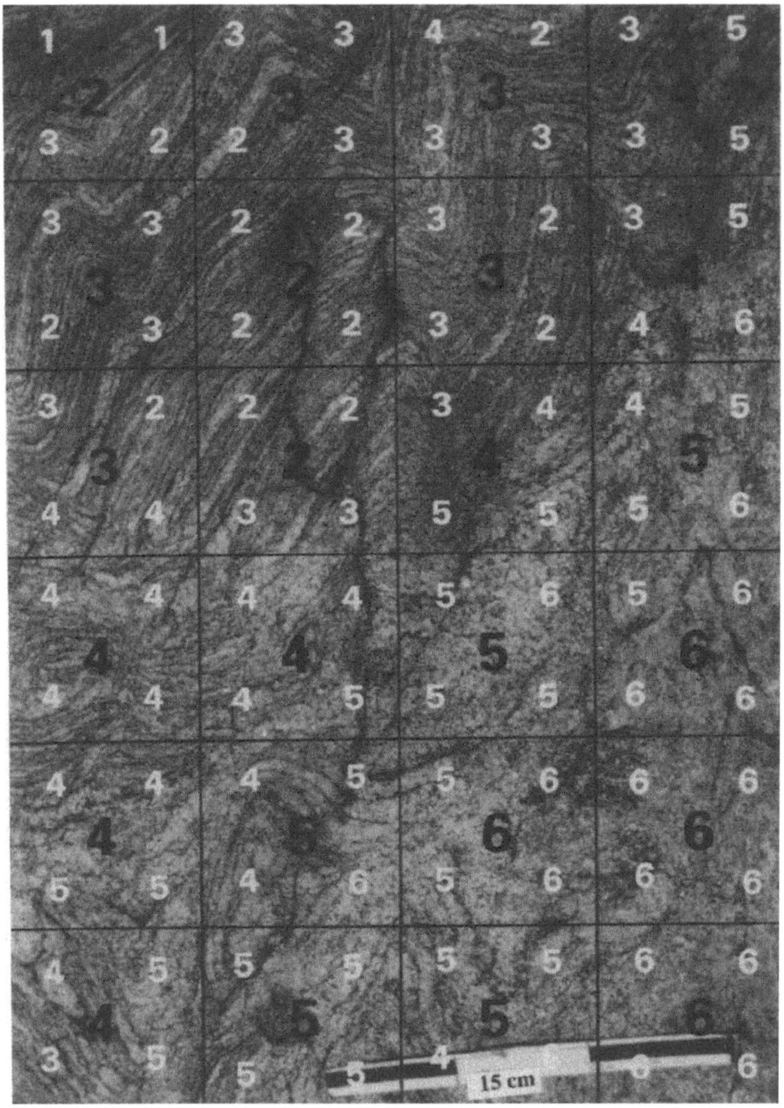

Figure 3 Application of textural classes to migmatites. The small-scale mapping of textures illustrates the variation of migmatitic textures in a hand sample. Black numbers in the center of each cell indicate the overall textural type for the 7 by 7 centimeter grid. Smaller white numbers indicate that the exact geometry may be further subdivided on a finer scale. The same mapping techniques can be used to describe the geometry of migmatites on a larger scale and to understand the distribution of texturally distinct migmatites on a crustal-scale.

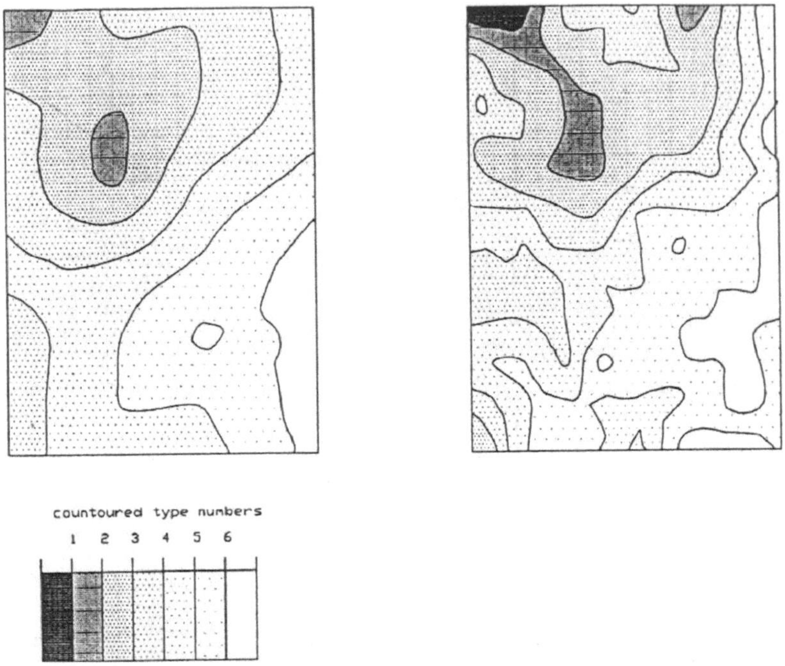

Figure 4 Same migmatite of Figure 3 contoured on the base of textural type numbers. The contour lines adequately reproduce the gradational evolution of textures as seen in the hand sample of Figure 3. The resolution of small variations decreases with the scale of mapping. Larger cell sizes on the other hand show the large-scale geometry of migmatites.

The mapping of textures at different scales indicates the possibility of self-similarity of migmatitic structures on even larger scales. Such a relationship could be used to support the prediction and extrapolation of structural associations in poorly exposed migmatite terrains. The contour diagrams of Fig.4 demonstrate how mapping results may then be recalculated and displayed on a computer. The textural type numbers 1 to 6 have been used as data input in the interpolation program. The diagrams show the results for both scales of resolution. Mapping on the next larger scale would result in a subdivision of Fig.3 into a small-scale folded (type 3 texture) upper half and an irregular to nebultic (type 4/type 5 texture) lower half of the migmatite sample. It can easily be seen that the assignment of textural classes strongly depends on the scale of observation.

Moreover, it is important to understand that the contour lines of Fig.4 do not represent discrete textural boundaries which delineate two distinct types of textures. The distribution of contour lines depict the geometry of the existing textural gradients which can also be seen in the migmatite sample of Fig.3. It may be argued that numerical modelling techniques will always result in the calculation of gradients between individual data points and that sharp textural boundaries may not be resolved, simply due to the choice of mathematical interpolation algorithms. The textural gradients seen in the final three dimensional model may therefore be a result of the contouring. This objection can be rejected because detailed field observations clearly indicate that, except for the occurrence of discordant intrusions, the large-scale variations of migmatitic textures are gradational and rarely characterized by sharp textural boundaries. Therefore the interpolation between individual data points of migmatitic textures is in good agreement with the general observations in migmatite terrains.

In order to work out the large-scale geometry of a migmatitic crust we have mapped an area of 10 by 10 kilometers based on 200 by 200 meter cells by simply defining the overall textural type within each cell. Cell sizes have locally been enlarged in areas of poor outcrop or in a few places where the outcrops were inaccessible. The mapping results have then been used to create a complete three dimensional model (10 by 10 by 5 kilometers) by extrapolating surface data into the unexposed parts of the model area (see below). The three-dimensional visualization was done with "Earth Vision" release 1.1 by Dynamic Graphics Software. The program requires a set of x,y,z data and an associated property value. Again, we have used the observed or extrapolated textural type numbers 1 to 6 as the input property. The calculated model represents an expansion of Fig.4 into the third dimension so that contour lines are replaced by contoured isoshells of equal migmatitic textures in 3D-space. In comparison with Fig.4 the scale has been enlarged from centimeters to kilometers.

4. Extrapolation of surface data

A number of field observations indicate that geological reasonable extrapolation rules may be established for different parts of the model. The validity of the final three dimensional model depends on the reliability of these extrapolation methods. Due to its high topography of nearly 2 km and locally very steep relief the Argentera massif provides some naturally exposed cross-sections. In addition to these exposures the model is based on three important extrapolation methods:

1. Some areas in the model region are characterized by the intrusion of melt from deeper structural levels. These intrusions locally cross-cut earlier migmatitic textures although they are generally associated with large metasomatic and anatectic reaction zones in the surrounding migmatites and gneisses. The intrusions belong to the late stage thermal peak of the migmatitic event. The distribution of intrusions in the exposed parts of the model indicates that the volume of intrusive material increases with depth and that the geometries of the intrusions are irregular. These observations have been extrapolated into the 3D-model.
2. A more important extrapolation method results from detailed structural observations. It has been found that the distribution of early forming migmatitic textures (probably metamorphic migmatites) strongly depends on the large-scale geometries of existing fold structures. This relationship indicates that the early formation of stromatic, small-scale folded and locally irregular textures (type 2, type 3 and type 4 of Fig.2) is directly controlled by a strong regional syn-migmatitic deformation event. Small-scale folded textures often display a typical S, Z, M geometry depending on the position within the associated large-scale fold structure. Moreover, the small-scale folded textures dominate the cores of these folds, whereas the layered types (type 1 and type 2) dominate the limbs of the folds. A detailed structural survey of the study area and a reconstruction of the

overall fold geometry may be used for the extrapolation of migmatitic textures in some parts of the model. As the dimensions of the observed folds are in the order of kilometers, this basic structural relationship provides a powerful tool for the construction of a realistic large-scale three dimensional model.

3. Additional vertical control for the three-dimensional geometry can be obtained from greenschist facies mylonite zones which frequently displace different structural levels in the Argentera basement. The formation of fault zones is related to the intrusion of a late Variscan granite body and the final exhumation of the massif. Although most of the fault zones exhibit a negligible amount of vertical displacement (in the order of tens of meters) two larger fault zones with a significant amount of vertical displacement are present in the model region. The juxtaposition of different structural levels indicates that the existence of type 4 and type 5 textures decreases upwards into higher structural levels. This implies that most of the irregular and nebulitic textures are in fact related to the typical metasomatic reaction zones of the above mentioned syn-migmatitic intrusions. This relationship is important for the modelling of eroded higher structural levels.

The use of extrapolation rules is complicated by the fact that two deformation events have been recognized during the migmatitic stage. Moreover, the amount of late metasomatic and anatectic overprint is uncertain for the unexposed deeper parts of the model. It is therefore not possible to model the real distribution of textures within the Argentera Massif except for those areas with very good surface control. However, the resulting model and its geometric characteristics are based on the exact structural and thermal history of the Argentera basement in addition to those scaling parameters which have been obtained from continuous outcrops.

Despite these restrictions to the validity of the presented three-dimensional structural model we consider it to be a close approximation to the real complexity of a migmatitic crust. Modification of the model by increasing the amount of intrusive material or by changing the expected distribution of textures in some parts of the model does not really influence the general conclusion which can be drawn from the overall geometric properties of the migmatites.

5. Geometric results and seismic properties

A comparison of Fig.3 with the final model in Fig.5 helps to show the general structural style of the model. The migmatite in Fig.3 is characterized by a folded layering which in places had gradually developed into a more homogeneous texture. Our structural work in the Argentera Massif has shown that an older metamorphic layering is distributed in large kilometer-scale folds and that the amount of intrusive material and the associated metasomatic overprint increases with depth. The irregular spacing of contoured isoshells in Fig.5 indicates that the geometry of textural gradients (strong or weak) is very similar to the migmatite in Fig.3. It seems that migmatites as viewed on a crustal-scale show very similar geometric properties to those observed in individual outcrops. As the large-scale geometry of the model should be used to discuss the seismic properties of migmatites, the geometric parameters must somehow be related to the potential reflectivity.

Figure 5 Three dimensional model of a migmatite terrain. The large-scale distribition of migmatitic textures is based on the mapping of textures on a 200 by 200 meter grid. The crustal-scale distribution of texturally distinct migmatites is very similar to the small-scale textural variation in Figure 3. Dark and light colors represent lower and higher textural type numbers, respectively.

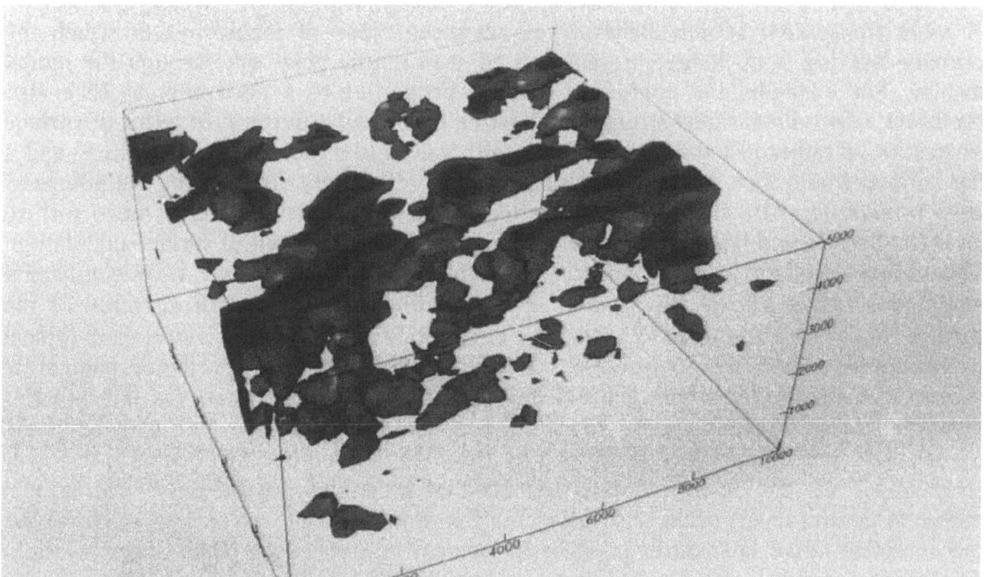

Figure 6 Distribution and geometry of non-migmatitic layering (type 1 to type 2 textures) without the strongly migmatized crust of the model. The model allows individual display of contoured isoshells of equal migmatitic textures which helps to show the overall complex geometry.

The classification scheme of textures (Fig.2) is based on the degree of layering. Despite the importance of crustal-scale mylonite zones, the generation of deep seismic reflections has been attributed to result from alternating layers of different seismic velocities (eg. Fountain and Salisbury, 1981; Hurich and Smithson, 1987; Rey, 1993). This origin of deep seismic reflectivity also depends on the existence of laterally extended layering. The large-scale structural model of Fig.5 clearly indicates that regional migmatization phenomena tend to destroy any type of pre-existing layering on a crustal scale. It suggests that the potential of reflectivity in migmatites depends on the size and orientation of the remaining non- or poorly migmatized layers (type 1 and locally type 2 textures). However, it must be demonstrated that the newly forming migmatites are distributed gradually over large distances and that they generally lack seismically detectable anisotropies.

The advantage of three dimensional contouring techniques is that it is possible to analyse the geometric results in a number of different displays. It is possible to display the isoshells of only one type of texture or a group of textures separately (see Fig.6) which helps to show the existing geometric effects. Fig.6 shows the distribution of the preserved layering within the model cube. Non-migmatitic layers are distributed in large elliptical lenses and boudins or in arrays of chained bulges which range in the order of hundreds of meters up to several kilometers long. The long axis of the lenses and bulges appear to be parallel to the orientation of the preserved layers. The size and shape of non-migmatized rock portions imply that in the case of a subhorizontal orientation some of the layers are likely to produce short isolated reflectors. In the Argentera basement the remaining layering has been folded and is preserved only in a very steep orientation. Seismic properties are therefore also controled by accompanying deformation events which in turn control the formation of migmatitic textures.

A more transparent seismic behavior of advanced types of migmatites in which the primary layering is no longer present can be seen in any plane cut through the model display. For example, the contoured textural properties of a horizontal xy-slice also represent a two-dimensional strongly undulating topographic surface. In terms of surface expression, a rather low topographic level with values less than 2 (type 2 texture) and a flat surface would then indicate a high potential of reflectivity disregarding the effects of layer orientation. Also the potential of reflectivity decreases with a higher mean surface level (higher textural type numbers) and with an increasing amount of surface undulation. The surface undulation may be quantified by defining the characteristic wavelengths and amplitudes between local topographic lows and highs in any chosen direction of the surface. In the case of our model cube the corresponding xy-sliceing would yield typical wavelenghths which are comparable with the extent of the preserved non-migmatitic layers. The amplitudes which express the diversity of textural properties indicate this diversity strongly depend on the structural position of the xy-slice with respect to the model. The mean amplitudes expressed by the contoured properties decrease in lower structural levels due to an increasing amount of intrusions but they also decrease in higher structural levels because of an increasing amount of preserved layers. However, the complete range of textures is generally present in any chosen plane of observation. The exact geometry and the distribution of migmatitic textures may also be seen by cutting cross-sections through the model display. Due to the undulating nature of textural properties in 3D-space and the basic fact that the equivalent surface expressions of any contoured isoshell do not really represent a discrete textural boundary (see above) it becomes quite obvious that the migmatitic model cube of Fig.5 has a strong tendency toward seismic transparency. The model may be used for the interpretation of seismic images because the resulting reflectivity patterns may be correlated with the geometry and distribution of non-migmatitic layering. The type and geometric properties of reflectivity may thus be seen to be characteristic of certain orogenic processes.

6. Discussion and Conclusions

Due to its nearly complete exposure and high topography the internal architecture of a typical Variscan migmatite terrain has been mapped and modeled based on textural criteria. The resulting three-dimensional model allows some general conclusions for the seismic properties of deeper continental crust. Seismic transparency in the central Variscan domain presumably results from the complex interaction of syn-migmatitic deformation processes and the intrusion of melt from deeper structural levels. This evolution results in:

-folding and disruption of potential seismic layers
-destruction of foliations and anisotropic layering
-textural and lithological homogenization
-formation of wide textural gradients and destruction of sharp textural or lithological boundaries.

The combined effects of deformation and migmatization on the seismic properties of continental crust and the evolution of reflectivity throughout time may also be illustrated by restoring the pre-migmatitic structural state of the Argentera basement. This can be achieved by extracting large bodies of anatectic melt intrusions from deeper structural levels and by unfolding large-scale tectonic structures which have accompanied the migmatization event. This hypothetical experiment would in fact result in a highly foliated sub-horizontal layering. Moreover, the lithological variations which have been found within the Argentera massif suggest that this layering could have been reflective prior to the migmatization event. Additional petrological data (von Raumer, 1993; Bogdanoff and Latouche, 1986) from the Argentera gneisses suggest that the restored sub-horizontal layering has been uplifted from a lower or at least deeper crustal position.This is supported by a strong ductile stretching event which has been preserved in non-migmatized metamorphic layers. After this early deformation event most gneisses developed high temperature equilibrium textures with polygonal grain boundries. PT-data from garnet amphibolites indicate 670-790° C/8-9 Kbar (Bogdanoff and Latouche, 1986) and 700-750° C/ 8-10 Kbar (Bierbrauer, unpubl. data) for this stage of evolution prior to the formation of migmatites.

The model of the present day large-scale geometry in the Argentera basement therefore clearly demonstrates that crustal reflectivity patterns evolve through an orogenic event. It suggests that crustal thickening and subsequent exhumation during the Variscan thermal evolution may have resulted in the formation of a seismically transparent upper crust and a highly reflective lower crust. This behavior is typically seen on reflection profiles of the Variscan domain (eg. Lüschen et al., 1987; see also Mooney and Meissner, 1992).

Moreover the structural model suggests that the occurrence of localized short reflectors in quasi transparent sections may be due to the preservation of local layering.

The processes of deformation and migmatization, which presumably control reflectivity, are related to the large-scale structural framework of the Variscan orogen. Therefore, the evolution of seismic properties strongly depends on the overall geometry of the collision zone, the structural position during the collision event, and the pre-existing structural and metamorphic history of the crustal area in question. The formation of larger greenschist facies mylonite zones which are also present in the Argentera basement may contribute to the final seismic appearance of the crust. Such fault zones contain strongly foliated phyllonitic mylonites which have been derived from the migmatites and which may be distributed in discrete zones up to several tens of meters thick. Some of the fault zones are accompanied by broader zones of proto-mylonites. The faults may be reflective due to the presence of fluids or distinct lithological and textural anisotropies within the fault zones and the surrounding migmatites.

The conclusions drawn from the Argentera massif are valid only for a migmatitic crust which has not experienced a strong post-migmatitic deformation event. Later deformations superposed on older migmatitic textures may result in the formation of new fabrics and textural anisotropies which in turn may result in new seismic reflectors. The structural data from the Argentera massif also indicate that the development of migmatitic textures is mainly related to associated larger syn-migmatitic deformation structures. In fact, folding of "leucosomes" does not necessarily indicate a post-migmatitic deformation event; and the apparent cross-cutting relationships of leucosomes often result from one or possibly two subsequent syn-migmatitic deformation events (see also Yardley 1978). Such textures may not indicate two migmatization events which are separated in time. These examples have been mentioned because the proposed mapping of high grade gneiss terrains based on textural criteria is also very useful to understanding the general structural evolution of a migmatitic crust.

Finally, a fractal distribution of migmatitic textures may be considered on a crustal-scale of resolution. The scale dependence for the application of textural type numbers has been pointed out earlier. The structural model in Fig.5. may be viewed as a characteristic large-scale intermediate crustal type 3 to 4. It may be subdivided into smaller cubes each representing a specific type of structure. This suggests that the classification scheme in Fig. 2 is also valid for distinguishing the different structural types of continental crust. However, the apparent self-similarity is limited to a vertical scale of several kilometers. Moreover, this self-similarity is not universal and different magnification factors have to be applied in order to reproduce a specific type of geometry. According to Mandelbrot (1985) this behavior is characteristic for self-affine fractals.

The importance of self-affine fractals in naturally deformed rocks has been demonstated by Schuman (1993) for the distribution of fold structures and strain in the central Appalachians. As pointed out by Schuman (1993) the self-affine fractal dimensions in rocks are rather variable because geological processes are generally being controlled by many different factors. This is also true for the formation of migmatites. The mapping of migmatitic textures in the Argentera Massif and our large-scale structural model suggest a self-affine fractal geometry for the distribution of migmatites, although this has not been demonstrated mathmatically. Possibly, large-scale blocks of type 1 to type 6 crust may be identified based on the occurence of characteristic reflection patterns. However, the model presented in this paper is only a first approach to integrate geometric parameters from migmatites to the understanding of crustal reflectivity patterns. It must be tested with models from other Variscan domains and compared with similar models from high grade gneiss terrains which have evolved under different tectono-metamorphic conditions. In the future an increasing number of such models may then be used to understand deep seismic reflection profiles in much more detail.

Acknowledgments

Supported by the German National Science Foundation DFG (ON 7/5-1, ON 7/5-2 and ON 7/5-3). We thank Bill Kidd, Ewald Lüschen and Klaus Weber for their helpfull reviews.

References

Amit, O. and Eyal, Y. (1976) The genesis of Wadi Magrish migmatites (N-E Sinai), Contrib. Mineral. Petrol., **59**, 95-110.

Ashworth, J.R. (1976) Migmatites, Blackie, Glasgow London. In: Ashworth, J.R. and Brown, M., eds. (1983) High-temperature Metamorphism and Crustal Anatexis, Unwin Hyman, London.

Berthelsen, A. (1960) Structural classification of gneisses as used in team work in SW Greenland. Report 21st Int. Geol. Congr., **26**, 69-71.

Bogdanoff, S. und Latouche, L. (1987) Evolution precoce du Massif de l'Argentera: appart des eclogites et des granulites, Geology Alpine, 63, 151-164.

Christensen, N.I. (1989) Reflectivity and seismic properties of the deep continental crust, J. Geophys. Res., **94**, 17,793-17,807.

Dohr, G.P. and Meissner, R. (1975) Deep reflections in Europe, Geophysics, **40**, 25-39.

Franke, W. (1989) Tectonostratigraphic units in the Variscan belt of central Europe, Geol. Soc. Am. Spec. Pap., 230.

Fountain, D.M. and Salisbury, M.H. (1981) Exposed cross-sections through the continental crust: Implications for crustal structure, petrology and evolution, Earth Planet. Sci. Lett., **56**, 263-277.

Hale, L.D. and Thompson, G.A. (1982) The seismic reflection character of of continental Mohorovicic discontinuity, J. Geophys. Res., **87**, 4525-4635.

Hurich, C.A. and Smithson, S.B. (1987) Compositional variation and the origin of deep crustal reflections, Earth Planet. Sci. Lett., **85**, 416-426.

Hopgood, A.M., Bowes, D.R. and Addison, J. (1976) Structural development of migmatites near Skåldö, southwest Finnland, Bull. Geol. Soc. Finnland, **48**, 43-62.

Hopgood, A.M., Andrew, St. and Bowes, D.R. (1978) Neosomes of polyphase agmatites as time-markers in complexly deformed migmatites, Geologische Rundschau, **67**, 313-330.

Johannes, W. and Gupta, L.N. (1982) Origin and development of a stromatic migmatite, Contrib. Mineral. Petrol., **79**, 114-123.

Lüschen, E. et al. (1987) Near-vertical and wide-angle seismic surveys in the Black Forest, SW Germany, J. Geophys., **62**, 1-30.

Mandelbrot, B.B. (1985) Self-affine fractals and fractal dimension, Physica Scripta, **32**, 257-260.

Matte, P. (1986) Tectonics and plate tectionics model for the Variscan belt of Europe. Tectonophysics, **126**, 329-374.

Matthews, D.H., and Cheadle, M.J. (1986) Deep reflections from Caladonides and the Variscan from Britain and comparison with the Himalayas. In: M. Barazangi and L.D. Brown (eds.), Reflection Seismology: A Global Perspective. Am. Geophys. Union, Geodyn. Ser., **13**, 5-19.

McLellan, E.L. (1989) Sequential formation of subsolidus and anatectic migmatites in response to thermal evolution, eastern Scotland, Journal of Geology, **97**, 165-182.

Mehnert, K.R. (1968) Migmatites and the origin of granitic rocks, Elsevier Publishing Company, Amsterdam London New York.

Mehnert, K.R. and Büsch, W. (1982) The initial stage of migmatite formation, N. Jb. Miner. Abh., **145**, 211-238.

Mooney, W.D. and Meissner, R. (1992) Multi-genetic origin of crustal reflectivity: A review of seismic reflection profiling on the continental lower crust and moho. In: D.M. Fountain, R. Arculus and R.W. Kay (eds.), Continental Lower Crust, Elsevier, Amsterdam London New York Tokyo.

O'Brien, P.J. (1989) The petrology of retrograded eclogites of the Oberpfalz Forest, northeastern Bavaria, West Germany, Tectonophysics, **157**, 195-212.

O'Brien, P.J., Röhr, C., Okrusch, M. and Patzak, M. (1992) Eclogite facies relics and multistage breakdown in metabasites of the KTB pilot hole, NE Bavaria: implications for the tectonometamorphic evolution of the NW Bohemian Massif, Contrib. Mineral. Petrol, **112**, 261-278.

Schuman, W. (1993) Fractal strain distribution and ist implication for cross-section balancing, J. Struc. Geol., **15**, 1497-1507.

Pin, C. and Vielzeuf, D. (1983) Granulites and related rocks in Variscan median Europe: a dualistic interpretation. Tectonophysics, **93**, 47-74.

Rey, P. (1993) Seismic and tectono-metamorphic characters of the lower continental crust in Phanerozoic areas: A consequence of post-thickening extension, Tectonics, **12**, 580-590.

von Raumer,J.F. (1974) Zur Metamorphose amphibolitischer Gesteine im Altkristallin des Mont-Blanc- und Aiguilles-Rouges-Massivs, Schweiz. Min. Petr. Mitt., **54**, 471-488.

von Raumer, J.F (1988) Caledonian-Variscan structures in the Alps - an introduction, Schweiz. Mineral. Petrogr. Mitt., **68**, 291-299.

von Raumer, J.F., Ménot, R.P., Abrecht, J. and Biino, G. (1993) The pre-alpine evolution of the external massifs, in J.F. von Raumer and F. Neubauer (eds.), Pre-Mesozoic Geology in the Alps, Springer-Verlag, Berlin Heidelberg.

Wimmenauer, W. and Stenger, R. (1989) Acid and intermediate HP metamorphic rocks in the Schwarzwald (Federal Republic of Germany), Tectonophysics, **157**, 109-116.

Yardley, B.W.D. (1978) Genesis of the Skagit Gneiss migmatites, Washington and the distinction between possible mechanisms of migmatization, Bull. Geol. Soc. Am., **89**, 941-951.

SEISMIC DATA OF THE CARPATHIAN FOREDEEP BASEMENT (ROMANIA)

CAMELIA DIACONESCU, V.RAILEANU, M.DIACONESCU, F.RADULESCU, A.POMPILIAN, M. BITER
National Institute for Earth Physics
P.O.Box MG-2, Bucharest-Magurele, ROMANIA

Abstract

The Carpathian foredeep is a molasse-bearing depression that formed in front of the folded eastern and southern Carpathians in the Late Miocene. Deep seismic reflection/ refraction profiles as well as gravity and thermal studies carried out in the foredeep area of the southern and eastern Carpathians provide information on the crustal structure of the study region. The variability in the reflectivity pattern and crustal thickness shown by the different sectors of the Carpathian foredeep are due to differences in the structure and lithology as well differences in crustal age (Klemperer, 1987; Wever et al., 1987).

In western part of the southern Carpathian foredeep reflectivity increases with depth and Moho is deliniated by strong reflections at 9-11.5 s two-way travel time (TWT) (30 - 33 km depth). The eastern side of the southern Carpathian foredeep shows a decreasing reflectivity with depth, the crustal base being interpreted at 12-13 s TWT (40-45 km depth) (Raileanu et al., 1994).

The eastern Carpathian foredeep is characterized by an almost transparent upper crust and a layered lower crust down to 13 s TWT (40-45 km depth). The greatest thickness of foredeep rocks is in the eastern Carpathian arc bend (Focsani depression), where Neogene rocks are 8 to 10 km thick.

O. Oncken and C. Janssen (eds.), Basement Tectonics 11, 125–140.

1. Introduction

The Carpathian foredeep external to the outer southern and eastern Carpathians is considered to be a typical molasse bearing foredeep that formed coeval with thrusting in the adjacent mountain belt (Royden and Karner, 1984). It is a transition zone between the Carpathians and foreland units and formed as a result of the Savian and Styrian tectonic movements (Burchfiel and Royden, 1982).

The Carpathian foredeep contains mostly Neogene sediments largely derived from the Carpathians, but also with significant contributions of detritus from the more external East - European (Moldavian) and Moesian platforms. Rock types within the foredeep include molassic conglomerates, sandstones and shales and local important evaporite deposits. Folds and small thrusts within the Neogene, Pliocene and even Pleistocene rocks are complicated by salt diapirs which have a variety of complex structural forms (Patrut et al., 1973). Many of these folds contain important oil reserves (Paraschiv, 1979; Burchfiel and Royden, 1982). The foredeep ranges in width from only a few kilometers (~10 km) in the northern part of the eastern Carpathians to more than 100 km at the southern Carpathian bend. The greatest thickness of the foredeep sediments is south of the eastern Carpathian bend (Focsani depression), where the Neogene rocks are probably 8 to 10 km thick. Recently this area has been subsiding at a rate of up to 3mm/yr. (Popescu and Dragoescu, 1986).

This paper synthesizes previous seismic studies carried out in this zone(Radulescu et al., 1977;Radulescu, 1981;Raileanu et al., 1994;Diaconescu M. et al., in press) creating an integrate interpretation of the data. The knowledge of the crustal structure of the Carpathian foredeep in Romania is mainly based on seismic (near-vertical reflection profiling and refraction methods) gravity and thermal data (Andreescu et al., 1989; Andreescu, 1993; Demetrescu and Andreescu, 1994;Diaconescu C. et al., in press).

2. Geological setting

Structurally, the Carpathian foredeep forms a wedge of clastic rocks and evaporites that thickens toward the Carpathians. The inner part of the foredeep has been involved in the folding and thrusting of the foreland belt.

The Carpathian foredeep was deformed during three different orogenic phases. The first stage ended in the Early-Middle Miocene with the uplift of the Carpathian orogen and overthrusting of the flysch zone onto the foredeep deposits.

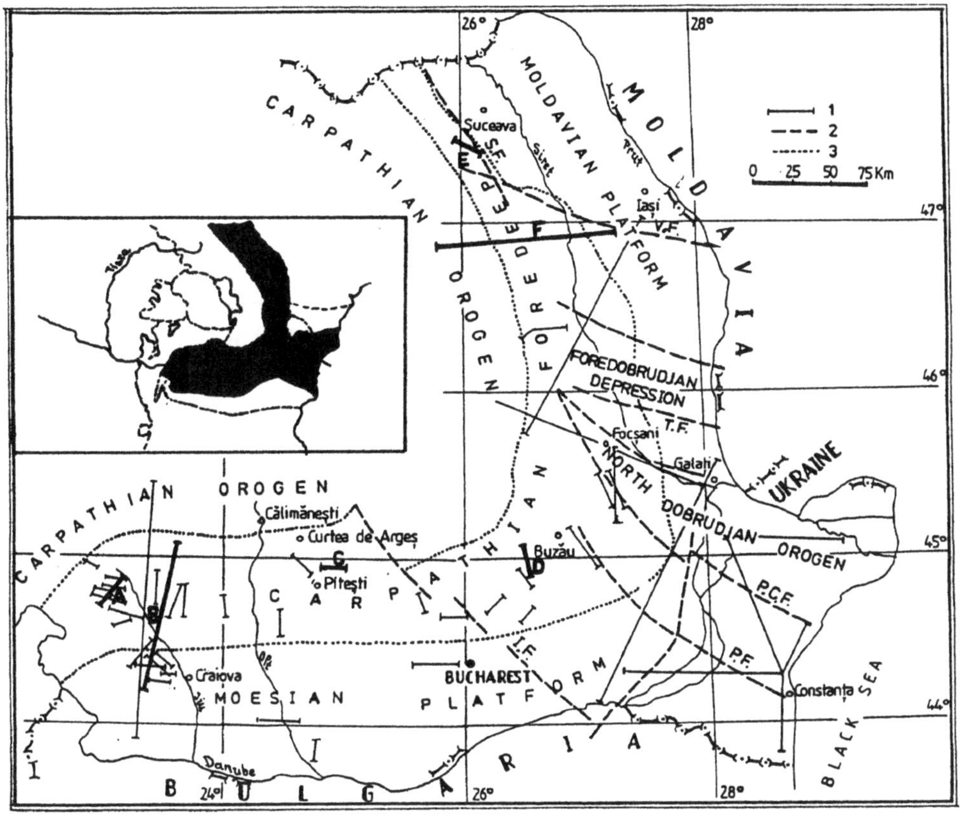

Figure 1. Location map of deep seismic reflection and refraction lines in the Carpathian foredeep and neighbouring areas. 1 - seismic line; 2 - fault; 3 - limit between tectonic units; I.F.-Intramoesian fault; P.F.-Palazu fault; P.C.F.- Peceneaga-Camena fault; T.F.- Trotus Fault; S.F.-Solca fault; V.F.-Vaslui fault.

At the same time, the foredeep deposits together with the Palaeogene flysch were strongly folded. During the Late Miocene (Moldavian orogenic phase) the folded molasse zone overlapped the foreland areas along the Pericarpathian fault. During the Vallachian phase (Neogene-Pleistocene) salt deposits were folded as diapir folds (Mutihac, 1990). The sedimentary basin of the Carpathian foredeep became a separate crustal block in the Early Miocene as a result of the uplift of the Carpathian external flysch.

West of the Intramoesian fault (Fig. 1), the crystalline basement of the Carpathian foredeep is composed of Moesian type basement up to the South Calimanesti fracture, and of folded Hercynian elements of the southern Carpathians (Danubian autochthon)(Visarion et al., 1984). In this area the Moesian-type basement is made up of mesometamorphic schists, that underwent retrograde metamorphis during the Late Protozoic age (0.6-1.0 Ga)(Mutihac, 1990). Granitic, granodioritic and gabbroic Hercynian intrusions pierce the basement (Paraschiv, 1979). The folded Hercynian elements of the Danubian autochthon are composed of crystalline schists with granitic intrusions of pre-Alpine age and a sedimentary cover of pre-Tertiary age .

Between the Intramoesian fault and Peceneaga-Camena fault (Fig. 1), the crystalline basement of the Carpathian foredeep can be subdivided into two different rock types:
1-Rocks metamorphosed to the greenschist facies during the Cadomian period. They crop out in the central Dobrudja and extend west of the Danube between the Peceneaga-Camena and Palazu faults, where they are covered by sediments, (Fig. 1), (Sandulescu, 1984; Visarion et al., 1988 a)
2- Mesometamorphic crystalline schists of Karelian age that extend from the South Dobrudja westward, between the Palazu and Intramoesian faults (Visarion et al., 1988 a). The Intramoesian and Palazu faults are probably Cambrian, but are younger than the oldest strata of the Moesian sedimentary cover (Sandulescu, 1984). The Peceneaga-Camena fault is contemporaneous with the folding of the greenschists, but it was activated by the subsequent orogenic phases (Mutihac, 1990).

The Karelian crystalline basement was affected by tectonic and metamorphic processes at least one more during the recent Cadomian orogeny. The Variscan orogeny significantly affected the sedimentary cover, but had only minor effects on the basement (Sandulescu, 1984).

The greenschist facies basement extends northward, to the Vaslui fault (Fig. 1). North of this fault, the basement of the eastern Carpathian foredeep is composed of schists

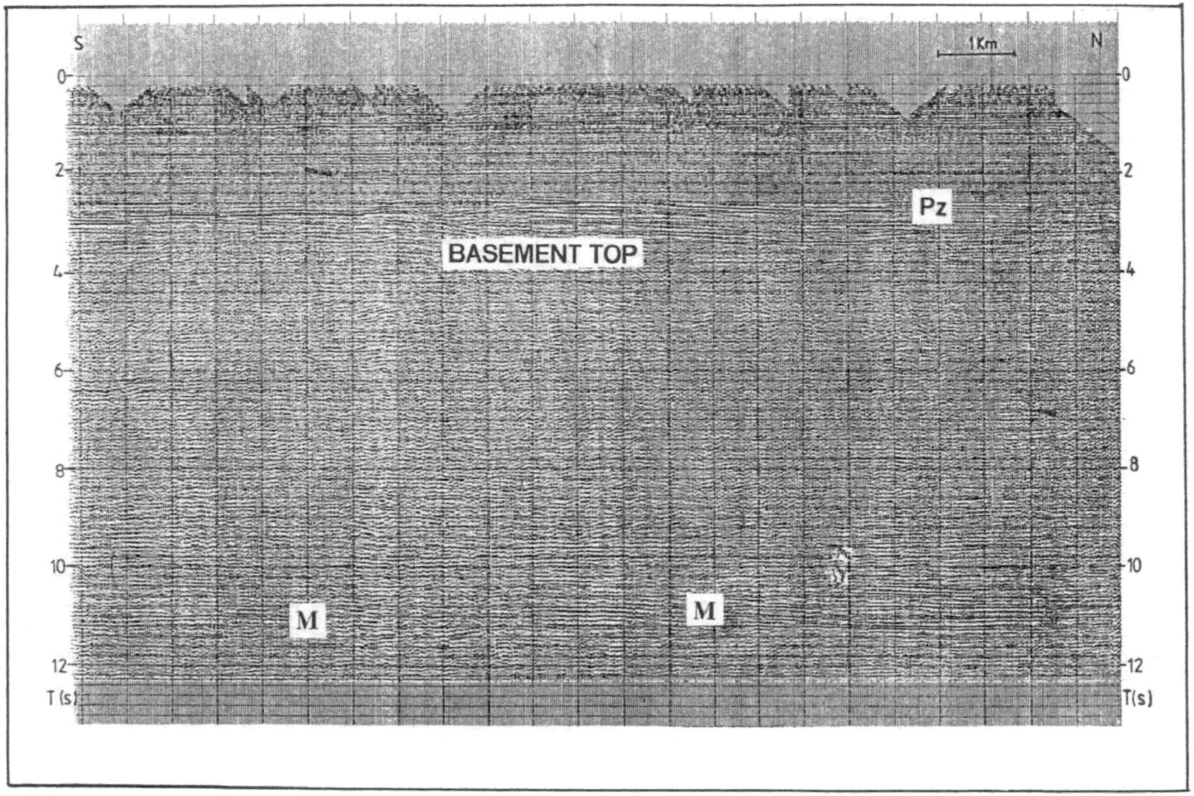

S
N
1Km
0
2
4
6
8
10
12
T(s)
Pz
BASEMENT TOP
M
M

of Baikalian age (North Dobrudjan type). In the East-European (Moldavian) platform the Baikalian basement on the west has been overthrust onto Middle Proterozoic basement to the east along the Solca fault (Visarion et al., 1988 b).

3. Geophysical data

Many deep seismic reflection and refraction lines were recorded in the Carpathian foredeep area, most of them in the westernmost zone of the southern Carpathian foredeep and in the eastern Carpathian arc bend region. Using seismic analyses together with other geophysical data (Radulescu et al., 1976; Cornea et al., 1981; Radulescu, 1981; Radulescu, 1988; Enescu et al., 1992), the crust has been divided in different domains, some of them being transitional boundaries.

Line A (see Figure 1 for location) was recorded in the western part of the Carpathian foredeep, in the region with Moesian platform-type basement. The seismic section of this line (Fig. 2) shows a highly reflective sedimentary cover of which the Palaeozoic top (Devonian) is the most prominent horizon, at 2.8-3.0 s two-way travel time (TWT). The top of the crystalline basement does not generate prominent reflections, but a weak reflective horizon at ~ 3.8 s TWT(7-8 km depth), is interpreted to be the base of the sedimentary cover. The depth to the basement was evaluated based on drilling and refraction seismic data (Radulescu et al., 1977; Radulescu et al., 1984). The upper crust seems to be transparent (4.0-8.0 s TWT), possibly as a result of the homogeneity found with granitic intrusions (Brown, 1991), while the upper crustal lacks reflectivity, the lower crust has a strongly layered character (7- 11.5 s TWT). Such a spatial correlation is often cited as evidence for the extensional origin of the deep reflectivity (Brown, 1991). Mafic sills emplaced during the collision delamination could be an important contribution to the enhanced lower crustal reflectivity (Nelson, 1991). The Moho seems to be a transition zone between the crust and mantle (9.0 - 11.5 s TWT), is deliniated by two bands of reflections (Raileanu et al., 1994).

Figure 3 shows a crustal cross section of the line B (see Figure 1 for location) in the same region that was interpreted from reflection/refraction, gravity, geological and thermal data (Andreescu et al., 1989; Diaconescu C. et al., in press). The reflectivity pattern of this region is similar to the reflectivity of Variscan and Caledonian Europe indicating a stable crust (Nelson, 1991) that has not been deformed since Alpine orogeny. The magmatic intrusions that accompanied the orogen were horizontally dispersed as a result of the reduced viscosity caused by the relative high (400 - 500^{0} C) temperatures in the lower crust (Fig. 3) (Wever et al., 1987). The inferred temperature distribution in the crust is based on observe heat flow data (Diaconescu C. et al., in press).

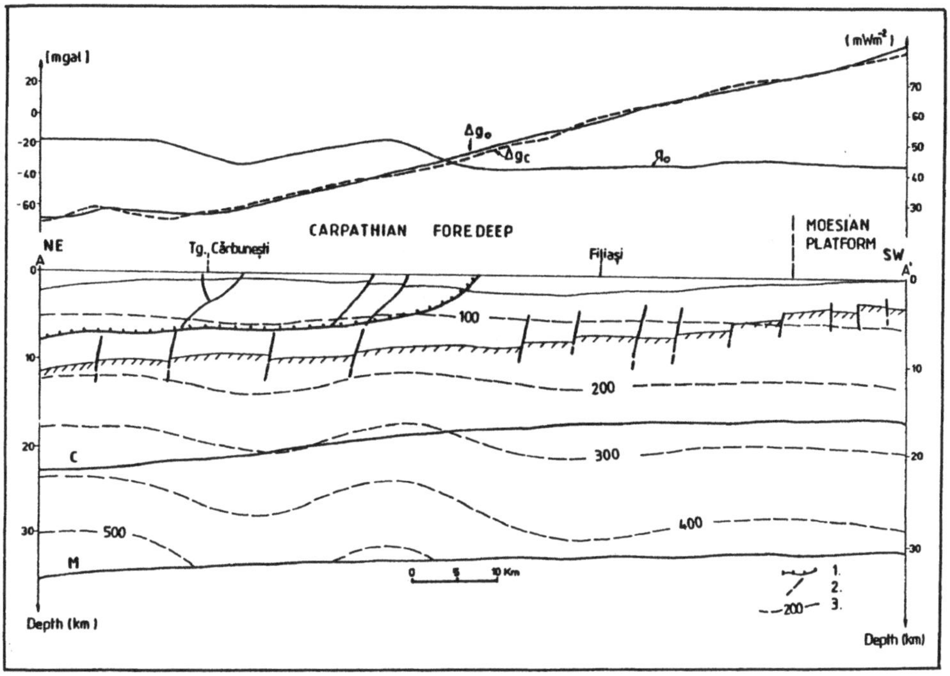

Figure 3. Depth section of line B in Figure 1. Δg_o - observed Bouguer anomaly; Δg_c - computed Bouguer anomaly; q_o - surface heat flow; C-Conrad and M-Moho discontinuities

Line C (Fig. 4) was recorded in the central Carpathian foredeep (Fig. 1). The basement top is interpreted to be at ~ 5.5-6.0 s TWT (~ 10 km depth), close to the bottom of the reflective sedimentary cover. The crust shows a decreasing reflectivity with depth, the upper crust

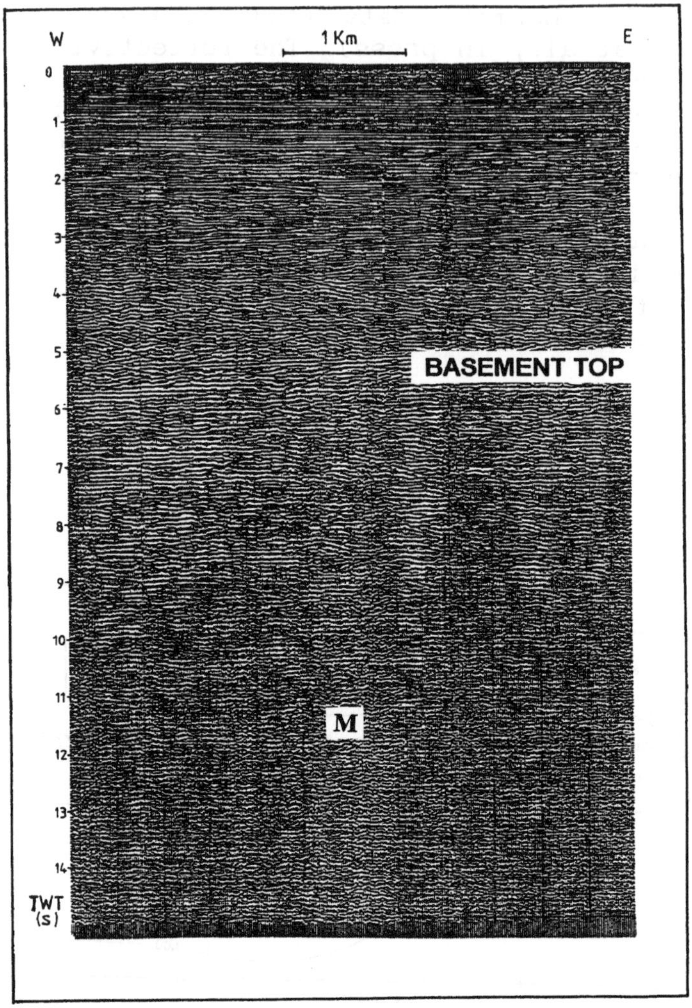

Figure 4. Seismic section of line C (see Figure 1 for location); M-Moho.

contains zones with strong reflections that can be correlated over long distances, alternating with zones void of reflections. At the same time, some short and diverging reflections like seismic 'crocodiles' in southern Germany (Sadowiak et al., 1991) occur in the middle crust (6.0-8.0 s TWT). They are found in compressional belts and may represents different interwedged rocks (Price, 1986).

The base of the crust is characterized by a paucity of reflections, is interpreted to transition to mantle material at ~ 11.6-12.0 s TWT (~36 km depth), with a mean crustal velocity of 6.2 km/s (Enescu et al., 1992).

The line in Figure 5 was recorded in the arc bend zone of the eastern Carpathian foredeep (line D in Figure 1). The top of the basement is interpreted to be at ~ 5.0 s TWT, at the base of the sedimentary cover. Reflections from 6.2-6.8 s TWT delineate the transition zone between the upper and lower crust. The Moho is deliniated by the strong nearly continous reflections from 12.8-13.0 s TWT. The seismic section of line D shows many diffractions mainly produced by fault zones. The reflections dip slightly from S to N, following the deepening of the Moesian platform deposits toward the orogen.

Fig. 6 is a density crustal section across the

Figure 5. Seismic section of line D (see Figure 1 for location). M-Moho.

eastern Carpathian foredeep along the international profile XI, (Radulescu et al., 1976; Cornea et al., 1981). It shows a gravity low value in the deepest part of the Carpathian foredeep (Focsani depression), that contains Neogene rocks that are 8 to 10 km thick .

Line E (see Figure 1 for location) is situated in the northern part of the eastern Carpathian foredeep, in the transition zone from the Moldavian platform to the eastern Carpathians. The Moldavian platform is the southwestern part of the east European platform, and it consolidated in the Middle-Proterozoic time. Line E crosses the Solca

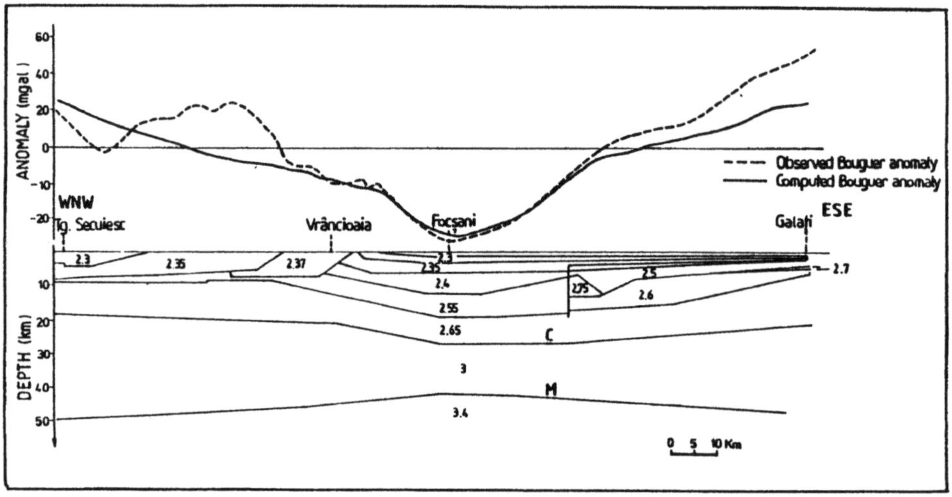

Figure 6. Density crustal section across the Carpathian foredeep arc bend zone (see line XI in Figure 1).

fault that separates two different types of basement: the Moldavian platform-type basement to the east and the Caledonian-Hercynian basement to the west. The Solca fault also separates two different reflectivity patterns (Fig. 7). The surface of the basement is interpreted to be at 2.8 s TWT (~ 6.5 km depth) in the southeastern side of the seismic section and it sinks slightly to the northwest up to the Solca fault. East of the Solca fault there is an increasing reflectivity with depth. The lower crust seems to be layered (between 10.0 and 13.0 s TWT), suggesting a ductile medium. West of the Solca fault, the seismic section has many diffractions, and the reflections are difficult to correlate. The Moho is inferred to be at ~ 42 km depth (Raileanu et al., 1994).

A crustal model across the eastern Carpathian foredeep as interpreted from a refraction line (line F in Figure 1) recorded in the transition area between the Moldavian platform to the Carpathian orogen is presented in Figure 8. It depicts the main crustal domains, as follows: the sedimentary/basement boundary (9-11 km depth), the

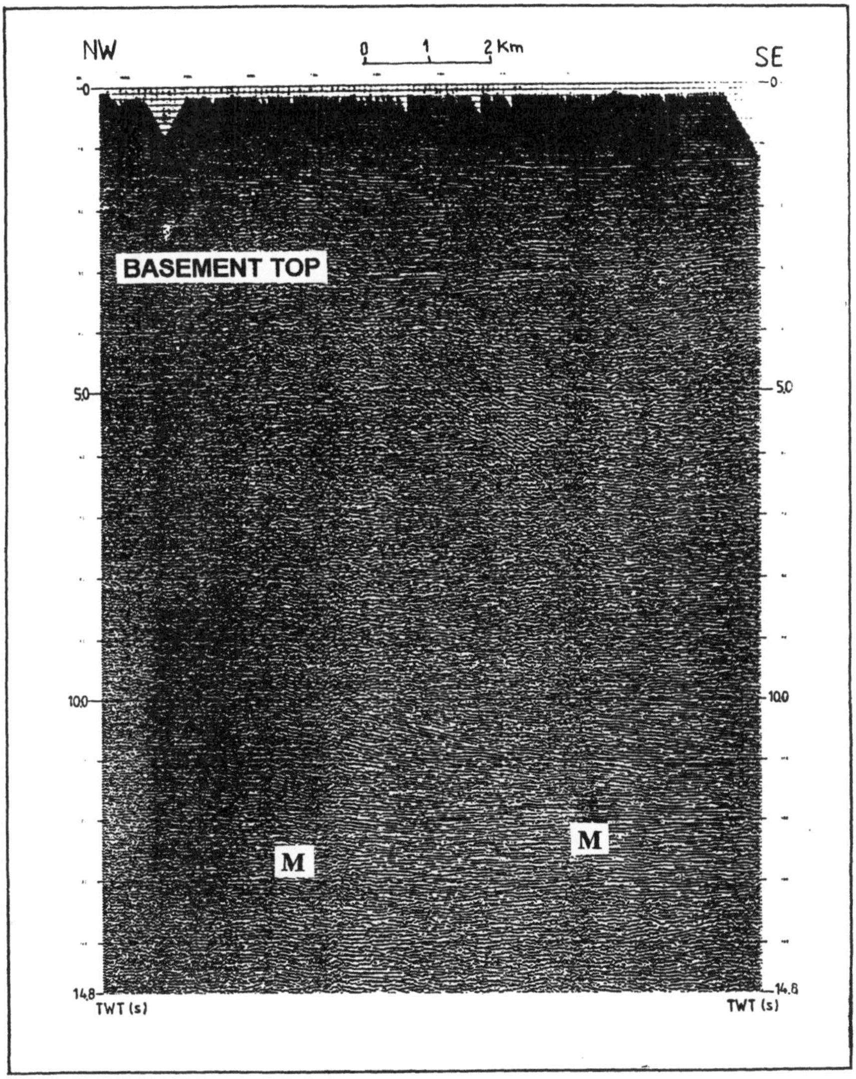

Figure 7. Seismic section of line E (see Figure 1 for location); M-Moho.

upper/lower crust limit (19-22 km depth), the base of the crust (43-50 km depth) and the base of the lithosphere (75-78 km depth) (Pompilian et al., 1993).

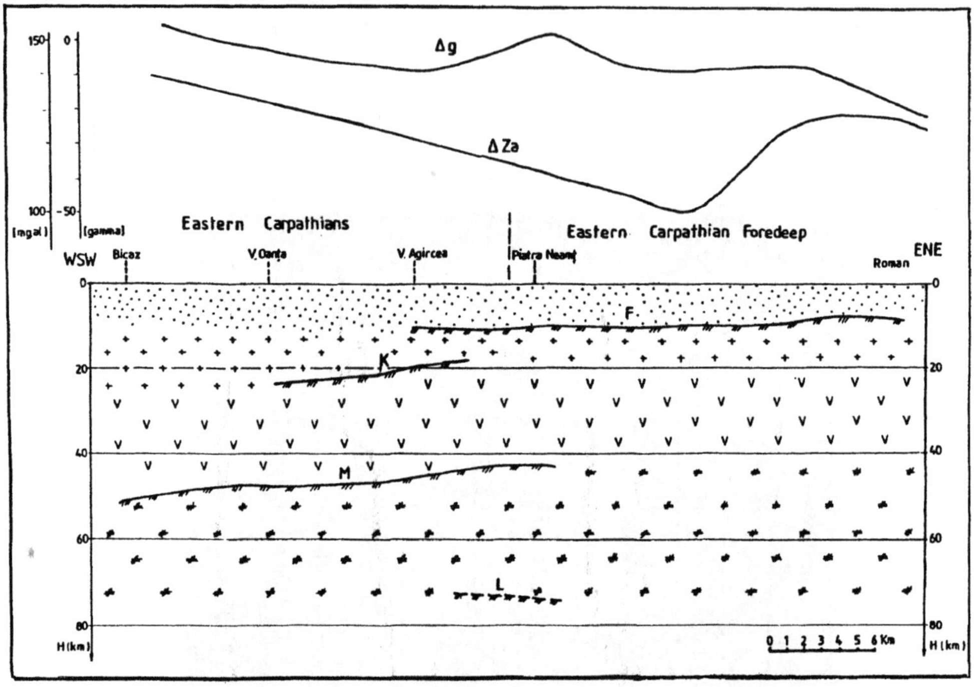

Figure 8. Crustal cross section across the Eastern Carpathian foredeep (F in Figure 1).F-sedimentary/basement boundary; K-Conrad discontinuity; M-Moho; L-lithosphere/ asthenosphere boundary; ☐ — sedimentary cover; ☐ — upper crust; ☐-lower crust; ☐-subcrustal lithosphere

4. Conclusions

The western part of the southern Carpathian foredeep shows a relatively transparent upper and middle crust with some diffractions and/or some short and dipping reflections that may indicate a brittle medium (Fig. 2). The reflectivity increases with depth and a crust-mantle transition zone with distinct and prominent reflections (at 9.5-11.5 s TWT) has been identified. The temperatures calculated (using heat flow data) for the Moho boundary in this region have values around 400-500^{0} C , suggesting thermal conditions that would induce the development of layering of the lowermost crust. This suggests that the Carpathian foredeep basement was reworked during the Alpine orogeny.

In central and eastern parts of the southern Carpathian foredeep the upper crust shows a pronounced reflectivity that decreases with depth, however the Moho is difficult to distinguish. This reflectivity pattern agrees with the relatively low values of the surface heat flow (35-40 mWm^{-2}) and the low temperatures calculated for the Moho (Andreescu, 1993). The crust thickens from west (~30-33 km) to east and northeast (~40-45 km). Moreover, the lower crust exhibits a thickening in this direction relative to the upper crust, which may explain its lower heat flow values (Demetrescu and Andreescu, 1994).

The eastern Carpathian foredeep formed as a basin with molasse sediments between Early Miocene and Late Pliocene. Although the reflectivity of the eastern Carpathian foredeep crust was poorly investigated, it does show some distinct characteristics. The northern side of the eastern Carpathian foredeep exhibits increasing reflectivity with depth, and the lower crust seems to be layered, suggesting a ductile medium (Wever et al., 1987). The crustal reflectivity of the eastern Carpathian foredeep which is 40 to 42 km thick is consistent with the $400^{o}C$ temperatures at a depth of 20 km (Demetrescu and Andreescu, 1994) that are present in this area.

5. References

Andreescu, M., Burst, D., Demetrescu, C., Ene, M. and Polonic, G. (1989) On the thermal regime of the Moesian platform and Getic depression, *Tectonophysics* **164**, 281-286.

Andreescu, M. (1993) The inverse approach to a geothermal profile of the central Moesian platform, Romania, *Tectonophysics* **224**, 435-442.

Brown, L. D. (1991) A new map of crustal 'terranes' in the United States from COCORP deep seismic reflection profiling, *Geophys. J. Int.* **105**, 3-13.

Burchfiel, B.C. and Royden, L. (1982) Carpathian foreland fold and thrust belt and its relation to Pannonian and other basins, *AAPG Bull.* **66(9)**, 1179-1195.

Cornea, I., Radulescu, F., Pompilian, A. and Sova, A. (1981) Deep seismic sounding in Romania, *PAGEOPH* **119**, 1144 -

1156.

Demetrescu, C. and Andreescu, M. (1994) On the thermal regime of some tectonic units in a continental collision environment in Romania, *Tectonophysics* **230**, 265-276.

Diaconescu, C., Raileanu, V., Radulescu, F., Diaconescu, M., Andreescu, M., Demetrescu, C. and Enescu, D. (in press) Complex geophysical study in the western part of the Moesian platform and Carpathian foredeep, *Rev. Roum. Geol. Geophys. Geogr., Ser. Geophys.*

Diaconescu, M., Radulescu, F., Diaconescu, C., Biter, M. and Pompilian, A. (in press). Crustal information from seismic and gravity data in the Moesian platform and Carpathian foredeep, *Rom. Rep. in Phys.*

Enescu,D., Danchiv,D. and Bala, A. (1992) Lithosphere structure in Romania II.Thickness of Earth's crust. Depth-dependent propagation velocity curves for the P and S waves, *St. Cerc. Geol., Geofiz., Geogr., Ser. Geofizica* **30**, 3-19.

Klemperer, S. L. (1987) A relation between continental heat flow and the seismic reflectivity of the lower crust, *J. Geophys.* **61**, 1-11.

Mutihac,V. (1990) Geological structure of the Romanian territory, Tech. Publ. House, Bucharest, 442 pp (in Romanian).

Nelson, K. D. (1991) A unified view of craton evolution motivated by recent deep seismic reflection and refraction results, *Geophys. J. Int.* **105**, 25-35.

Paraschiv, D. (1979) The Moesian platform and its hydrocarbon deposits, Acad. Rom., Bucharest, 195 pp (in Romanian)

Patrut, I., Paraschiv, D. and Dicea, O. (1973) Considerations on the forming of diapir structures in Romania, *Petrol si Gaze* **24(9)**, 533-542 (in Romanian)

Pompilian, A., Radulescu, F., Diaconescu, M., Biter, M. and Bala, A. (1993) Refraction seismic data in the eastern side of Romania, *Rom. Rep. in Phys.* **45(7-8)**, 613-621.

Popescu, M. and Dragoescu, I. (1986) Map of the recent crustal movements in Romania. Similarities and differences, *Geodynamics* **8**, 123-136.

Price, R. A. (1986) The southeastern Canadian Cordillera: thrust faulting, tectonic wedging, and delamination of the lithosphere, *J. Struct. Geol.* **8**, 239-254.

Radulescu,D.,P., Cornea,I., Sandulescu,M., Constantinescu,

P., Radulescu, F. and Pompilian, A. (1976) Structure de la croute terrestre en Roumanie. Essai d'interpretation *des* etude sismique profunde. *An. Inst. Geol. Geof.* **L(50)**, 5-36 (in French)

Radulescu, F., Pompilian, A. and Sova, A. (1977) Deep seismic sounding in Romania (Craiova-Petrosani area), *Rev. Roum. Geol., Geophys., Geogr., Ser. Geophys.* **21(2)**, 273-276.

Radulescu, F. (1981) Crustal seismic studies in Romania, *Rev. Roum. Geol., Geophys., Geogr., Ser. Geophys.* **25**, 57-74.

Radulescu, F., Raileanu, V. and Cornea, I. (1984) Contributions of the reflection seismic method to the crustal structure deciphering, *St. Cerc. Geol., Geofiz., Geogr., Ser. Geofizica* **22**, 11-17 (in Romanian).

Radulescu, F. (1988) Seismic models of the crustal structure in Romania, *Rev. Roum. Geol., Geophys., Geogr., Ser. Geophys* **32**, 13-17 (in Romanian).

Raileanu, V., Diaconescu, C. and Radulescu, F. (1994) Characteristics of Romanian lithosphere from deep seismic reflection profiling, *Tectonophysics* **239**, 165-185.

Royden, L. and Karner, G. D. (1984). Flexure of lithosphere beneath Apennine and Carpathian foredeep Basins: evidence for an insufficient topographic load, *AAPG Bulletin* **68(6)**, 704-712.

Sadowiak, P., Wever, Th. and Meissner, R. (1991) Deep seismic reflectivity patterns in specific tectonic units of western and central Europe, *Geophys. J. Int.* **105**, 45-54.

Sandulescu, M. (1984) Geotectonics of Romania, Techn. Publ. House, Bucharest, 336 pp. (in Romanian).

Visarion, M., Stanica, D. and Stanica, M. (1984) Deep structure of the region between the Motru Valley and Arges Valley (Romania) as shown by the results of geophysical surveys, *Acta Geodaet., Geophys. et Montanist. Hung.*, **19(1-2)**, 173-184.

Visarion, M., Sandulescu, M., Stanica, D. and Veliciu, S. (1988a) Contributions a la connaissance de la structure profonde de la Plate-forme Moesienne en Roumanie, *St. Teh. Econ.*, **D(15)**, 211-222.

Visarion, M., Sandulescu, M., Stanica, D. and Atanasiu, L., (1988b) An improved geotectonic model on the east Carpathians, *Rev. Roum. Geol., Geoph., Geogr., Ser. Geophys.*, **32**, 43-52.

Wever, T., Trappe, H. and Meissner, R. (1987) Possible relations between crustal reflectivity, crustal age, heat flow, and viscosity of the continents. *Ann. Geophys.*, **5B(3)**, 255-266.

PRE-PALEOZOIC CRUSTAL TECTONICS CONTROLLING PALEOZOIC AND TERTIARY METALLOGENIES IN THE CENTRAL REGION OF ARGENTINA

HUGO G.L.BASSI
CIRGEO
Ramirez de Velasco 847, 1414 Buenos Aires. Argentina.

ABSTRACT. The region comprises the provinces of Córdoba and San Luis where the crystalline basement crops out. Study of the distribution of tungsten and chalcofile ore deposits identified a 500 km long and 50 km wide belt where old and modern ore deposits coexist.

The NNE belt, which is accompanied by important Paleozoic granites and Tertiary volcanics, has strong Bouguer gravity anomalies. This belt includes a block of crust with anomalous thicknesses that practically contains the total of the Tertiary Pb, Zn and Au ore deposits At the margins of the block, where the crust has more typical thickness, tungsten Paleozoic deposits are present. Reactivated Pre-Paleozoic fractures systems control the whole of the magmato-metallogenic events that took place up to the present.

Introduction

The present paper addresses the magmatic and tectonic aspects of an identified metallogenic zone in the central region of Argentina that comprises the provinces of Córdoba and San Luis. It concerns about a NNE trend belt of about 500 km long and 50 km wide with selective distribution of Miocene epithermal ore deposits (Andean cycle) flanked on both sides by Paleozoic hypothermal ore deposits (Caledonian cycle).

Geological Framework

The mountains of both provinces form part of the Pampean ranges system, that include a crystalline basement occasionally covered by sediments and volcanics. In the 30,000 km2 region the basement is completely exposed except for 1,5 % of its total area which is covered by upper Tertiary volcanics.

The basement is composed of metamorphic stratified rocks and massive igneous rocks. The stratified rocks occupy 80 % of the area and primarily consist of high grade gneisses and schists with lesser amounts of quart-feldspathic veins, amphibolites, limestones and quartzites and lower grade metamorphic rocks which include phyllites and sericitic-chloritic schists.. There is also a small volume of non-metamorphic rocks such as orthoquartzites, graywackes, conglomerates and

O. Oncken and C. Janssen (eds.), Basement Tectonics 11, 141–147.

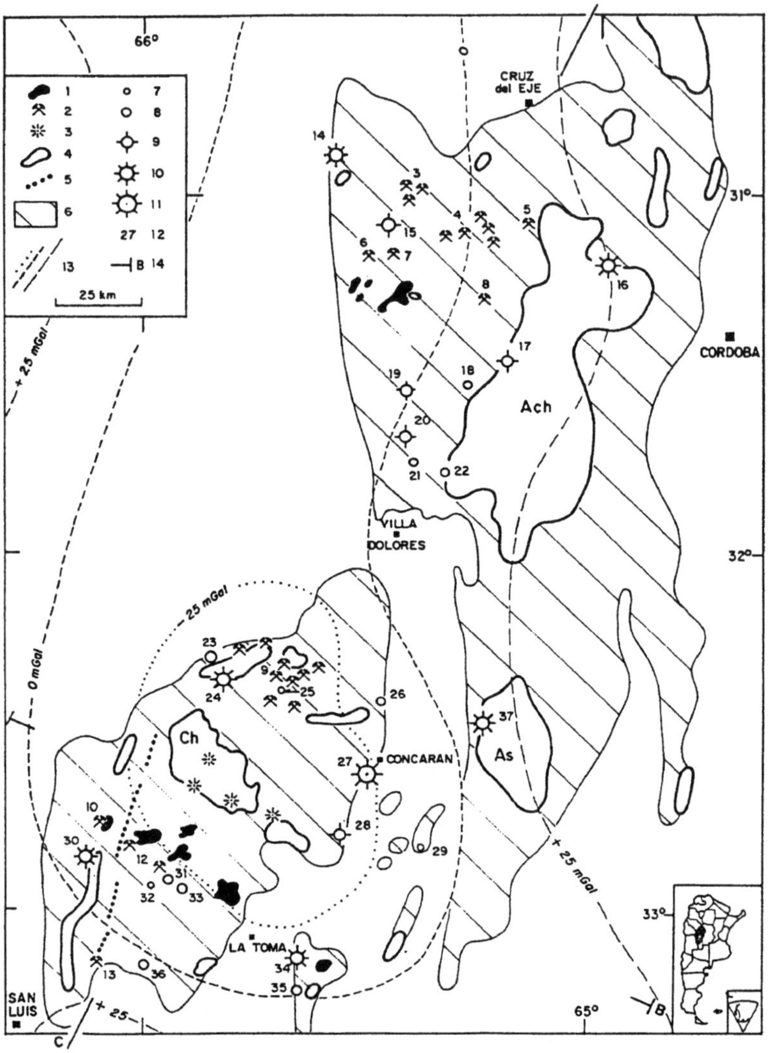

Figure. 1. Geological and mineral deposits map of the provinces of Córdoba and San Luis. Tertiary: 1, mediosilicic volcanics; 2, mineral deposits of the Andean Metallogeny (Au, Pb, Zn) . Mesozoic: 3, alkaline volcanics. Proterozoic-Paleozoic: 4, granitic intrusions (batholiths: Ach= Achala; As= Cerro Aspero; Ch= Las Chacras); 5, ultrabasic intrusive rocks; 6, mountainous areas (metamorphics); 7-11, tungsten deposits (Caledonian Metallogeny). Tons. of ore with 0.5/1 % WO_3 : 7, < 2,500 tons; 8, 2,500-30,000 tons; 9, 30,000-100,000 tons; 10, 100,000 to 300,000 tons; 11, over a million tons; 12, number of mine; 13, free air gravity isoanomalous contour curves ; 14, gravity profiles.(B and C)

sandstones (Lucero, 1979). Generally the schistosity of the basement rocks is parallels the original sedimentary bedding and strikes northnortheast and dips steeply. The metamorphic rocks of Proterozoic age (+- 640 Ma) have intercalated older amphibolites, up to 1400 Ma (Cingolani and Varela, 1975).

Four magmatic episodes underwent the basement: two intrusive and two effusive. The first episode was dominated by basic and ultrabasic rocks which are concordant bodies that in many instances are aligned with the NNE structural grain. The most important sequence of these rocks are in a 70 km long belt that lies along the western edge of the San Luis range (Fig. 1-5). The rocks range in composition between norites and perknites and are syntectonic intrusions with a Proterozoic shear belt that is about the same age as metamorphics.

The second and most extensive magmatic episode involved the emplacement of granitic intrusions. This episode occurred in three pulses during the Ordovician, Devonian and Carboniferous (Linares and Latorre, 1969). The area of the granitic outcrops occupies 20 % of the ranges (Fig. 1-4-). Despite the abundance of these intrusions only three of them are of batholithic proportions: the Achala batholith in the north; the Cerro Aspero batholith in the south and the Las Chacras batholith in the west. It is interesting to note that all of them are aligned along controlled the same trends which, as shown further on, are related to the crustal tectonics and metallogenies.The Achala batholith is elongated in a NNE trend, the Las Chacras batholit in a WNW direction and the Cerro Aspero batholit follows both directions.

During the third magmatic episode in the Cretaceous minor trachybasalt flows erupted from volcanic centers in eastern and southern Córdoba, outside the studied region, and in the Las Chacras granitic batholith in San Luis (Fig. 1-3-). It is interesting that these volcanic centers were aligned along NNE and WNW trends, similar to those of the granitic batholiths intruded two hundred million years earlier. The magma from which these volcanics were derived apparently came from the lower crust (Lopez and Sola, 1981) and may have originated in the mantle..

The fourth and last magmatic episode is represented by Tertiary intermediate volcanic rocks in San Luis and Córdoba and to a very minor extent in Santiago del Estero. In San Luis a succession of volcanic cones occur over a 60 km long WNW trend. Two hundred kilometers to the NNE, in Córdoba, there is a 25 km long belt of volcanic cones which, as in San Luis, also trend WNW. And fifty kilometers farther to the NNE, in Santiago del Estero, there are some intrusive bodies and isolated andesite dykes which might be related to the Córdoba and San Luis volcanic flows. This volcanism, which mostly took place during the Upper Miocene coeval with uplift of the Pampean Ranges, may have had its origins in andesitic magmas that were derived in the mantle connection but assimilated crustal rocks rich in potassium as they ascended (Gordillo and Linares, 1981).

In summary three magmatic episodes occurred in Córdoba and San Luis that are related to the base of the crust and possibly to the mantle. A fourth cycle is linked to the sial. The known metallogenesis is a consequence of some of these episodes; the intramagmatic metallogeny (Cr, Ni, Co) connected to the Precambrian ultrabasics; the hypothermal (W, Sn, Bi) to the Paleozoic granites and the epithermal one (Au, Ag, Pb, Zn, Cu, U, Mn and F) to the Tertiary intermediate volcanics.

From a tectonic point of view, the Pampean Ranges and, particularly those of Córdoba and San Luis, make up a block system of Tertiary age. In this system, the blocks are bounded by longitudinal NNE trending fractures and by transverse fractures that trend WNW. These fractures are thought to be older reactivated systems, that have been acting since early geological times. The Precambrian metamorphic structure as well as the basic contemporaneous intrusions follow the NNE trend.

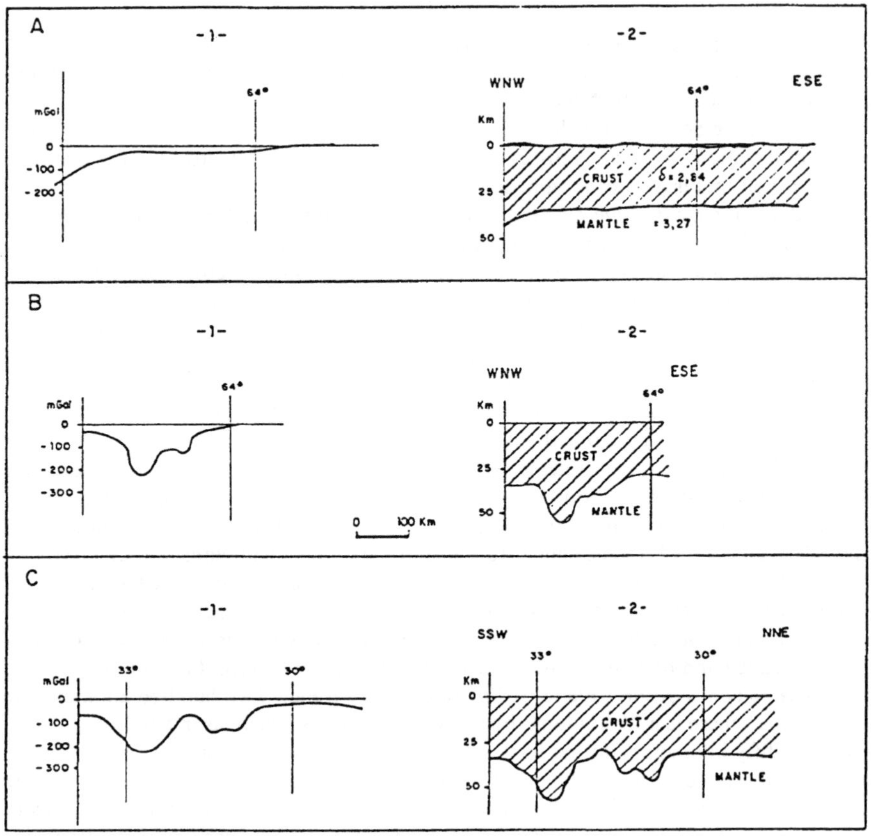

Figure 2. Gravity profiles and models of the crust. A, Transcontinental (San Luis-Cordoba section, Dragicevic, 1974). B, local, south of San Luis and Cordoba; C, local, San Luis-Cordoba-Santiago del Estero. 1, Bouguer anomalies (B and C, based on Browin et al., 1981); 2, models of crust, Pomposiello 1993, (see Fig. 1 for location).

These fractures were also reactivated during Pre-Carboniferous, Cretaceous and Tertiary orogenesis and even today are active faults (Amos et. al., 1981). The majority of the seismic epicenters of the region occur in a meridian trend alignment (Introcaso et. al., 1987) that coincides with the western border of the Andean Metallogenic Belt.

The NNE fracture system was apparently the dominant system in the tectonic history of the region. Although the transverse (WNW system) seems to be of major magnitude, it is more obscure and consequently there are few references about it in the literature. Thus, the WNW trending Carolina-Morro lineament (Bassi, 1948 and Pastore-Gonzalez, 1954) served as conduit for Tertiary volcanics in San Luis. The drainage of the Paleozoic peneplain as well as the Caledonian tungsten metallogeny in the Carolina and Morro districts (Bassi and Carrillo, 1991) are controlled by the WNW direction. The Carolina-Morro lineament projection to the WNW coincides with the Juan Fernandez ridge in the Pacific Ocean which is accompanied by a group of transcurrent faults in the oceanic crust (Howell et.al., 1985) and it is part of the structural discontinuity responsible for the segmentation of the Andean Orogen and as such has implications for the magmatism and metallogeny of the Cordillera (of Sillitoe, 1977).

GRAVITY INTERPRETATION

With the aim to confirm the crustal condition of those fractures some gravity values obtained over the region were examinated. In the Fig. 1 (-14-) free-air gravity isoanomalous curves outlined by Browin et al. (1981) can be seen. These curves show high and low gravity anomalies indicating important crustal disturbances for the central part of the Argentina territory. To interpret them they were transformed to Bouguer values with an average density of 2.84 gr/cm3 for the crust and 3. 27 gr/cm3 for the mantle (Pomposiello, 1993). In order to show the crust outside the study area, a sector of the transcontinental profile of Dragicevic (1974) has been reproduced. The A profile, situated few kilometers south of the San Luis Range, indicates a crust with a subhorizontal base at about 30 km depth (Fig. 2). The B profile, across the metallogenetic belt and already outlined in the San Luis Range, indicates that the base of the crust is at 30 km depth on both sides of the belt but its center the depth reaches is more than 50 km. In the C profile, which parallels the NNE trend metallogenetic belt, the crust is more than 50 km thick in the south in the San Luis and the center of the belt, under the Sierra de Pocho Norte.

GEOLOGICAL INTERPRETATION

Two 50 km thick segments of the crust coincide with volcanic centers and with the Tertiary epithermal metallogeny. In interpretating the B and C gravity profiles the block margins are delineated by deep crustal fractures (Fig. 3). Through these fractures which extend from the surface to the mantle the following have moved : 1) Proterozoic ultrabasic magma with the associated segregation ore deposits (Pt, Ni, Co) ; 2) Paleozoic granitic magmas and associated tungsten ore deposits ; 3) Cretaceous basic magmas ascending from the upper mantle or lower crust resulting in basalt-basanite volcanism with no mineralization and 4) Tertiary mediosilicic volcanics with the Au, Pb, Zn ore deposits.

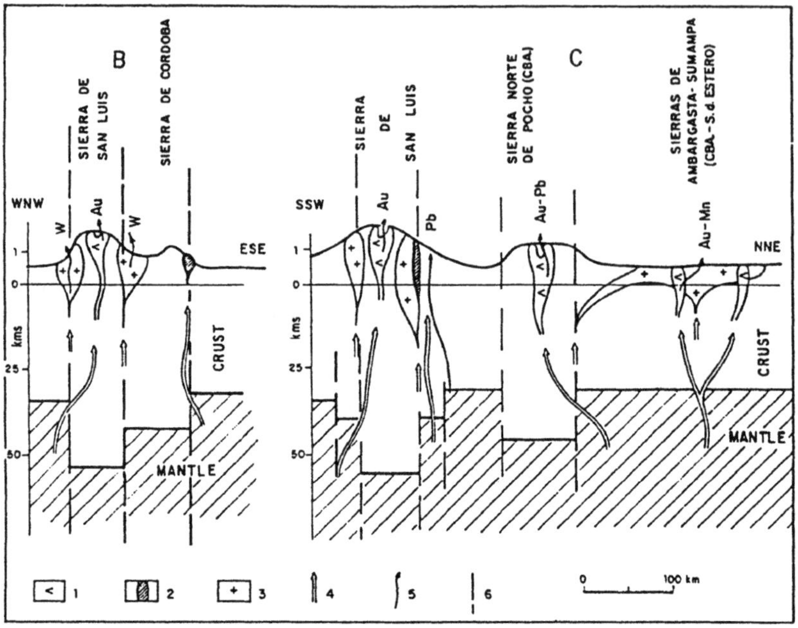

Figure 3. Tecto-magmato-metallogenic interpretation of the gravity anomalies in Cordoba and San Luis. B, Transverse Section; C, Longitudinal Section. Tertiary: 1, andesite-trachyandesite; ore deposits of the Andean Metallogeny (Au, Pb, Zn). 2, basic and ultrabasic rocks (Proterozoic) or trachybasalts (Cretaceous). Paleozoic: 3, granites; ore deposits of the Caledonian Metallogeny (W). 4, magmatic ways. 5, metallogenic fluids ways. 6, crustal fracturing.

Conclusion

The Proterozoic, Paleozoic, Mesozoic and Tertiary magmatism as well as the Caledonian and Andean metallogenies of the region follow well defined NNE and WNW trends, accomodating themselves to the Gondwanic Regmagenic Network (Bassi, 1988). The fracturing that controls such magmato-metallogenies and the block tectonics has a Pre-Paleozoic origin.

Acknowledgments

This paper has benefitted from three anonymous reviewers who provided some helpful comments which significantly improved the manuscript and the English version.

References

Amos, A. J., Calligari, R. H. and Siches, C. W. (1981) Active faults in the Argentine Republic (Spanish), *VII Congreso Geologico Argentino*, **II**, 235-242, Buenos Aires.
Bassi, H. G. L. (1948) Auriferous alluvial placers of the Carolina-Rio de la Carpa area , Prov. of San Luis (Spanish), *Revista de la Asociacion Geologica Argentina*, **III**-1, 5-51, Buenos Aires.
-------------- (1988) Hypotheses concerning a regmagenic network controlling metallogenic and other geologic events in the South American Austral Cone, *Geologische Rundschau*, **77**/2, 491-511, Stuttgart.
Bassi, H.G.L. and Carrillo, R. (1991) The regmagenic control in the geology and metallogenies of the El Morro district (Spanish), *Anales de la Academia Nacional de Ciencias Exactas, Fisicas y Naturales*, **42**, 183-194, Bs.As.
Browin, C., Warsi, W. and J. Millingan, (1981) *Free-air gravity anomaly atlas of the world*, Woods Hole Oceanographic Institution, Woods Hole, Massachusetts, 02543.
Cingolani, C. and Varela, R. (1975) Rb, Sr geochronology of the igneous and metamorphic rocks of the Chica and Grande Ranges of Cordoba, Argentine Republic (Spanish), *Act. 2° Congr. Iberoamer. de Geologia Economica*, **I**, 9-35, Buenos Aires.
Dragicevic, S. M. (1974) Gravimetric map of South America (Span), *Departamento de Geologia y Geofisica de la Universidad de Chile*, 1-10, Santiago.
Gordillo, C.E. and Linares, E. (1981) Geochronology and petrography of the tertiary volcanites of the Pocho department, Cordoba province (Spanish), *Rev. Asoc. Geolog. Arg.*, **36**, 4, 380-388.
Howell, D. G., Jones, D. L. and Schermer, E. R. (1985) *Preliminary tectono-stratigraphic terrane map of the Circum-Pacific region*, The Circum Pacific Council for Energy and Minero resources, American Asoc. Petrol. U.S.A.
Introcaso, A., Lion, A. and Ramos, V. A. (1987) Deep structure of the Cordoba Ranges (Spanish) *Rev. Asoc. Geolog. Argentina*, **42**, 1-2, 177-187.
Linares, E. and Latorre, C. (1969) K/A and Pb/alfa ages of granitic rocks of Cordoba and San Luis (Spanish), *IV Jornadas Geológicas Argentinas*, **II**, 195-204.
Lopez, M.G. and Sola, P. (1981) Alkaline volcanic manifestations of the environs of Las Chacras and the region of Villa Mercedes-Chajan, provinces of San Luis and Cordoba (Spanish), *VIII Congreso Geologico Argentino*, **IV**, 967-978, Buenos Aires.
Lucero M., H.N. (1979) *Pampean ranges of the north of Cordoba, south of Santiago del Estero, eastern border Catamarca and southeastern angle of Tucuman (Span.)*, in Geologia Reg. Arg., Acad. Nac. de Cs. de Cordoba, **I**, 293-347.
Pastore, F. and Gonzalez R. (1954) Description of the 23g sector of the argentinian geological map, San Fco., San Luis (Spanish), *Dirección Nacional de Mineria*, Boletin **80**, Buenos Aires.
Pomposiello, C. (1993) Interpretation of Bouguer anomalies and model of crust of Cordoba and San Luis, Contribution to this paper.
Sillitoe, R. H. (1974) Tectonic segmentation of the Andes: implications for magmatism and metallogenic, *Nature*, **250**, 542-545, U.S.A.

PRECAMBRIAN BASEMENT CONTROL ON 'PLAINS-TYPE FOLDS' (COMPACTIONAL FEATURES) IN THE MIDCONTINENT REGION, USA

D.F. MERRIAM
Kansas Geological Survey, The University of Kansas
Lawrence, KS 66047

ANDREA FÖRSTER
GeoForschungsZentrum Potsdam, Telegrafenberg A3
D-14473, Germany

Abstract

'Plains-type folds' were described in the early 1930s by petroleum geologists working in the Midcontinent, USA. The folds are characterized by small size, increased sharpness with depth, asymmetry, and association with faulting. These folds are ascribed to the compaction of sediments over crystalline, rigid basement, in this instance tilted Precambrian fault blocks. Initially, the 'plains-type folds' were formed when the present major structural features of the region were formed and adjustments continue today as recognized by neotectonics in the region. Development of these features in sediment veneers over the basement in the many cratonic environments of the world is of interest because locally they serve as traps for petroleum. Timing of development of the 'plains-type folds' determined when and if they were available for entrapment of petroleum as fluids moved through the system. This study indicates that there are two types of 'plains-type folds' in the Midcontinent: (1) the normal type developed mainly in the early Middle Pennsylvanian as a result of intraplate stresses caused by collision of the North and South American plates; and (2) those formed later as a result of other influences. By using a *structural-depth gradient,* which expresses the change in thickness of stratigraphic units on and off structure, it is possible to determine at which time(s) these features were affected by local or regional structural adjustments in the basement. By plotting the *structural gradient*, which is the change in structural closure with depth, the nature of the 'plains-type fold' can be recognized as being either (1) or (2). We propose that features exhibiting a low or no structural gradient (2), that is a similar structural closure with depth, and that are small and nearly circular in nature can be attributed to the emplacement of late Mesozoic or early Tertiary intrusive igneous plugs.

O. Oncken and C. Janssen (eds.), Basement Tectonics 11, 149–166.

1. Introduction

In the early 20th Century when oil had just been discovered in the Midcontinent and wells were being drilled with little or no information, it was important to have some guidelines for exploration. It was at this time that drillers followed 'trends' looking for anticlines, and the sophisticated exploration tools of the time were the planetable and alidade and the coredrill. From this work it was painfully obvious to these early workers that the anticlines mapped on the surface did not always coincide with the structure they were looking for at depth. They also noted that the structures, in addition to 'migrating,' were sharper and more pronounced in the older beds. Thus, the position and arrangement of the structural features suggested vertical rather than lateral or compressional movement. The origin of these features, later to be termed 'plains-type folds,' was best explained by differential vertical movement and draping of sedimentary units over rigid Precambrian basement faults blocks.

In the late teens and twenties of the 20th Century the description and origin of these features was the subject of many papers. Gardner in 1917 was perhaps the first to explain the vertical control of the features reasoning that lateral stresses could not have been transmitted through the sedimentary sequence for any distance and therefore could not account for the folds. Blackwelder (1920) likewise concluded that the folds were formed by a differential settling of the sediments over highs. Fath (1920) championed the transmission of stresses through the basement and intermittent movement of old faults of Precambrian origin. McCoy (1921) and Monnett (1922, 1927) explained the features by differential settling of the sediments. Powers (1922) discussed the possible origin of the structures and concluded that the compaction of sediments was the most important. Rubey (Rubey and Bass, 1925) concluded that block faulting of the rigid Precambrian took place differentially along preexistent lines of weakness.

In 1925, Powers summarized all the data on the structural geology in the Midcontinent and coined the term 'plains-type folding' for the small, irregularly distributed, periodically rejuvenated Midcontinent folds (Table 1). Several other papers at this time reiterated these ideas or described other areas. In 1932 Clark made a 'quantitative' study of the 'plains-type folds' and McCoy (1934) summarized all the theories on their origin. Rich (1935) thought the structural pattern, after the removal of regional dip, suggested readjustments in the fractured Precambrian basement. Little has been written on the subject per se since the mid-1930s but since that time much data have accumulated and subsequent studies have shown that the origin of these 'plains-type folds' are draping of sediment over a differentially rejuvenated fault-block pattern in the rigid Precambrian crystalline basement.

Table 1. Features of 'plains-type folds' as modified from Powers (1925).

* Structural noses are the dominant structure at shallow depth (ie in surface units)
* Most folds are small in areal extent and amount of structural closure
* Closure increases downward and the structure becomes more sharply defined
* The amount of closure is proportional to size in many structures
* The number of folds is inversely proportional to size and may be irregularly disposed

2. Plains-Type Folds

'Plains-type folds' are present worldwide in sedimentary covers of most continental shield areas. However, a search of the literature, using the key words 'plains-type folds,' revealed descriptions of them only in North America (where the term was proposed) and on the Arabian Shield (Sugden, 1962). Obviously, the term for these subtle anticlinal features, synonymous for *drape fold* or the Russian *placanticline* (Bates and Jackson, 1987), has not been universally accepted nor used much since it was proposed in the mid 1920s. Nevertheless, it is an apt term with genetic implications on the formational process for these types of features.

Although the structures usually develop slowly through time, abrupt changes in the pattern may reveal time(s) of more intense deformation. Sugden plotted the age of structural markers against the percentage of present structural development. This gave an indication of unconformities (or structural pattern changes) in the development of the features under study. Merriam (1963) devised a graphic method for determining the time of deformation by plotting the amount of closure against time and gave two examples from the Forest City Basin in northeastern Kansas.

2.1. DATA

Data for this study were obtained from the files of the Kansas Geological Survey in Lawrence. A combination of scout tops, drillers logs, sample logs, and wireline logs were used. Four oil fields were studied in detail - Slick-Carson, Coleman, Box, and Neosho Falls (Fig. 1). These fields were selected because the wells penetrated the older and deeper formations so enough data were available to apply the method.

Slick-Carson - This field was discovered in late 1924 in Cowley County and drilled in 1925 and 1926. The discovery well, Slick No. 1A Carson (sec. 19, T32S, R3E)[1], came in for about 900 barrels of oil per day (BOPB) from the 'Wilcox sand,' which is either remanent Simpson sand on the Arbuckle Group or the upper part of the Arbuckle, at about 3,450 ft[2]. Further development took place in the 1950s with discovery of oil in the shallow Pennsylvanian Bartlesville and Layton sands.

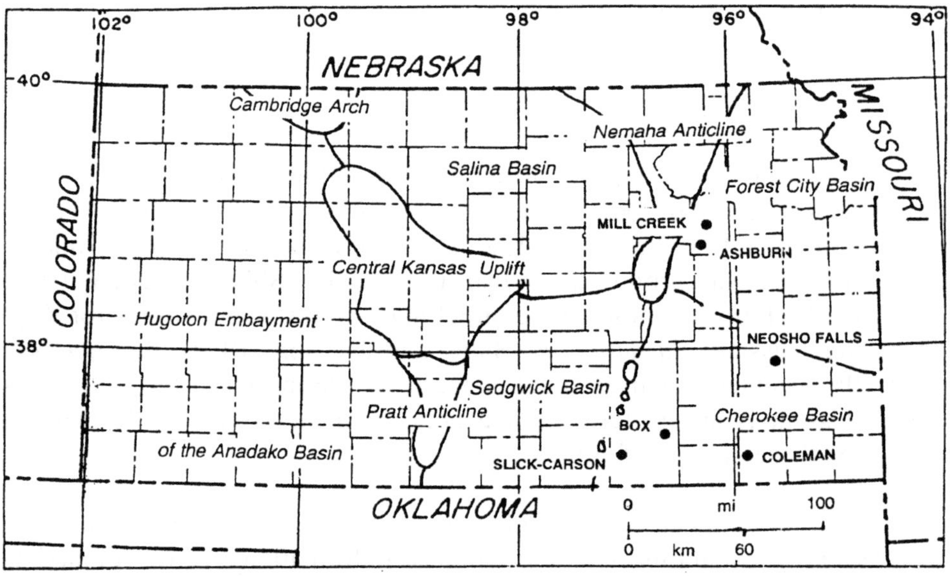

Figure 1. Index map of Kansas and part of adjacent states showing local anticlinal features discussed in text and major post-Mississippian pre-Middle Pennsylvanian structures of region (adapted from Merriam, 1963).

Coleman - The Coleman Field was discovered by the Danciger Bros. et al. No. 1 A.C. McGonigle well (NW SW NE sec. 33, T32S, R14E) in Montgomery County in 1924. The discovery well came in for 400 BOPD at a depth of about 1,700 ft. There was some additional drilling in the 1950s and Mississippian production was developed in the mid 1980s.

[1]This is the land coordinate position by section, township, and range
[2]3.28 feet = 1 meter; 0.62 mile = 1 kilometer

Box - The Box Field was opened in 1948 with discovery of oil (70 BOPD) in the Mississippian by the Wiedeman No. 1 Haydensmith well (SW NW NW sec. 28, T30S, R7E) in Cowley County. Further development took place in the late 1970s and early 1980s with Arbuckle production developed in the 1990s.

Neosho Falls Dome - Oil was discovered in Mississippian limestones on the Neosho Falls Dome in Woodson County in 1922. In 1928 the No. 8 Harris well (SWc sec. 8, T24S, R17E), encountered 'granite' beneath the Chattanooga Shale where the lower part of the Paleozoic section (Arbuckle and Reagan) was missing. The samples from which the presumably Precambrian rock type could be verified have since been lost. There is a possibility that the rock is a basic igneous rock (P. Berendsen, pers. comm., 1994). Subsequently there has been shallow gas production developed on the structure.

Contour maps were prepared for several stratigraphic horizons in each field. The amount of closure then was measured and plotted on a time/closure crossplot.

The post-Precambrian stratigraphic section in southeastern Kansas is short (Fig. 2). Lower Paleozoics, the Reagan Sandstone (Cambrian), Arbuckle Group (mostly carbonates; Cambro-Ordovician), and perhaps a very thin Simpson Group (sandstone; Ordovician) overlie the Precambrian crystalline rocks which consist mainly of epizonal and mesozonal granite with rhyolite and some metasediments. Overlying the lower Paleozoic strata is a thicker section of Mississippian carbonates (with a shale at the base) in turn overlain by a fairly thick section of Middle and Upper Pennsylvanian and Lower Permian clastics punctuated by thin carbonates in a cyclic arrangement. The Paleozoic strata have a thin cover of Tertiary and Quaternary deposits locally.

2.2. ANALYSIS

It was possible to prepare contour maps for the *Slick-Carson Field* on top of the 'Wilcox sand' (Simpson sand, Cambro-Ordovician), Mississippian, base of Kansas City Group (Missourian, Pennsylvanian), and base of the Oread Formation (Shawnee Group, Virgilian, Pennsylvanian) (Fig. 3). The crossplot is shown in Figure 4. There is a definite change in the structural pattern in the lower part of the Middle Pennsylvanian. This is the same stratigraphic place as determined by Merriam (1963) for a change in pattern in the Forest City Basin (Fig. 5). A cross section of the structure is shown in Figure 6.

For the *Coleman Field* contour maps were prepared on top of the Arbuckle (Cambro-Ordovician), Mississippian, base of Ft. Scott (Marmaton Group, Missourian, Pennsylvanian), and base of the Oread Formation (Shawnee Group,

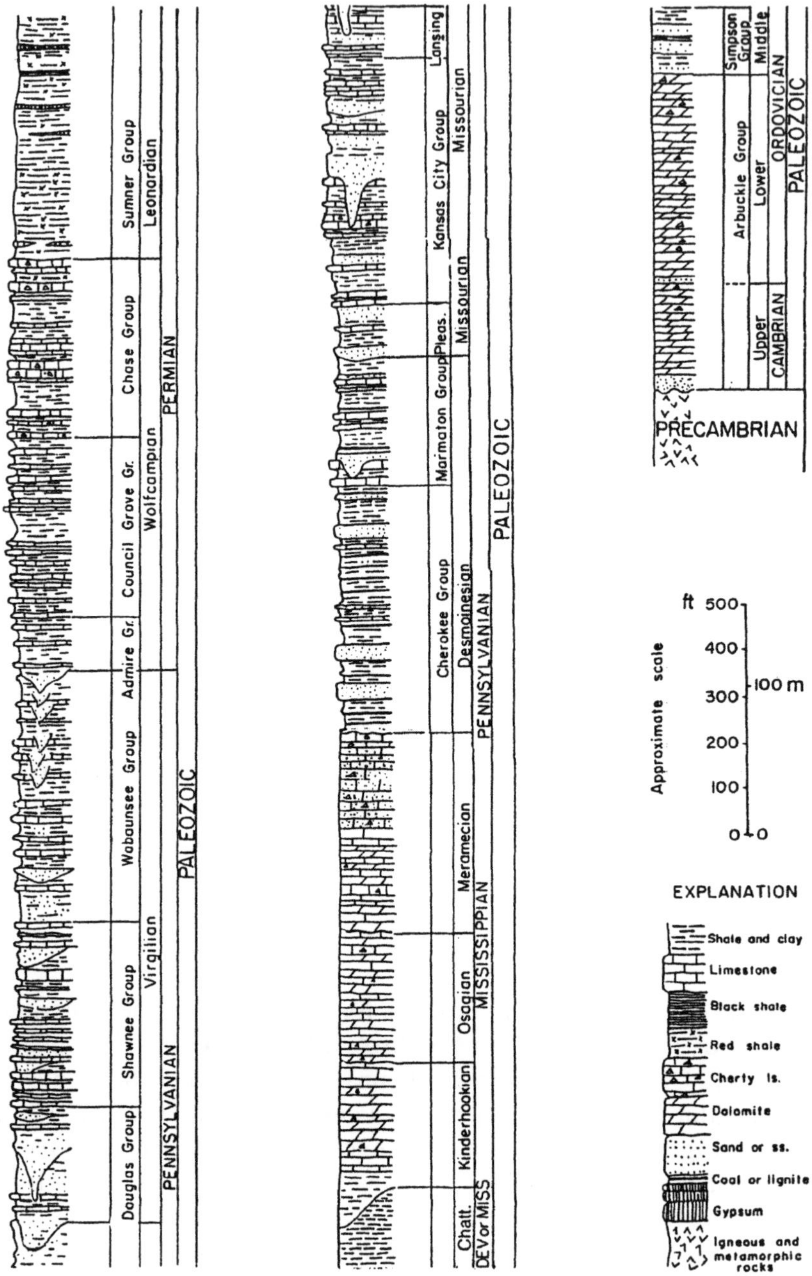

Figure 2. Generalized composite stratigraphic column for southeastern Kansas showing units mentioned in text and generalized lithology. Rocks older than Pennsylvanian are known from the subsurface only (adapted from Merriam, 1963)

Virgilian, Pennsylvanian). The crossplot for this field is similar to the Slick-Carson Field one in that there is an obvious change in structural pattern in the lower part of the Middle Pennsylvanian.

The crossplot of the *Box Field*, however, is different than Slick-Carson and Coleman (Fig. 7). The amount of closure on the Arbuckle (Cambro-Ordovician), Mississippian, base of the Kansas City Group (Missourian, Pennsylvanian), and base of the Oread (Shawnee Group, Virgilian, Pennsylvanian) is essentially the same on all horizons. One interpretation of this plot is that the structure was formed sometime after the Early Permian.

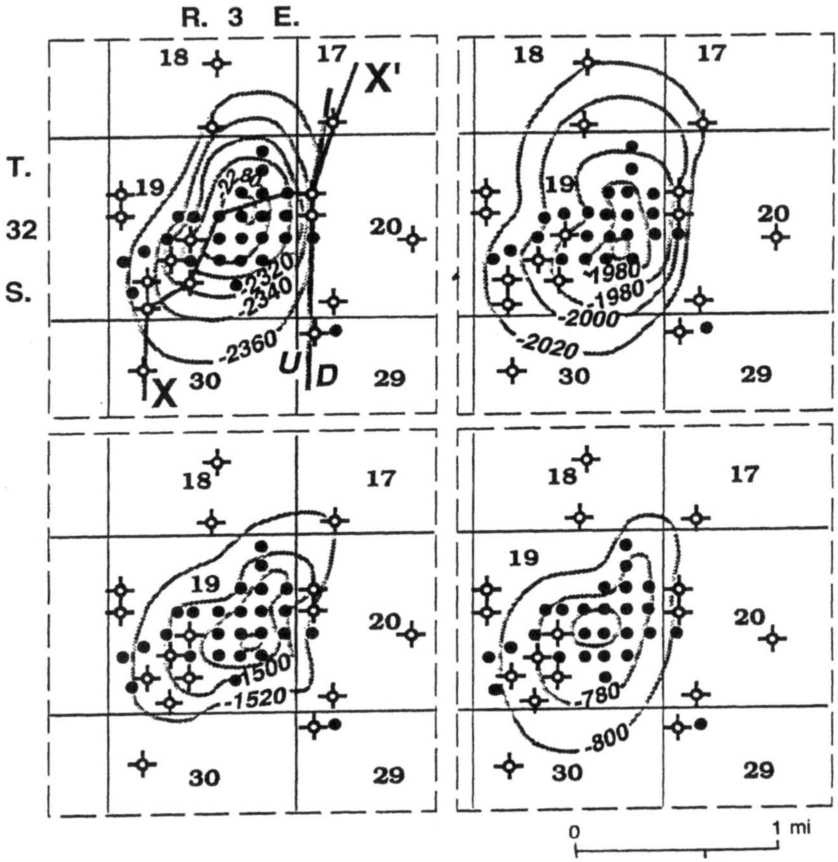

Figure 3. Configuration of contours on different stratigraphic horizons for Slick-Carson Field shown on (from oldest to youngest): A, top 'Wilcox sand' (Simpson sand); B, top Mississippian; C, base Kansas City Group; D, base Oread Formation (base Shawnee Group). (Location of field shown on Fig. 1.) Contour Interval (CI) = 20 ft (6.25 m).

Structure was mapped on four different stratigraphic horizons for the *Neosho Falls Dome*. Closure on top of the Mississippian, base of Ft. Scott (Marmaton Group), base of Kansas City, and top of Kansas City is essentially the same. The crossplot of closure versus geologic age however is similar to that for the Box Field and thus the igneous rock may be intrusive but rose only to the Chattanooga Shale. Unfortunately, not enough control was available to contour the structure on the older horizons, if indeed they are present. A cross section of the structure shows that crystalline rocks underlie the Chattanooga Shale at the crest of the dome.

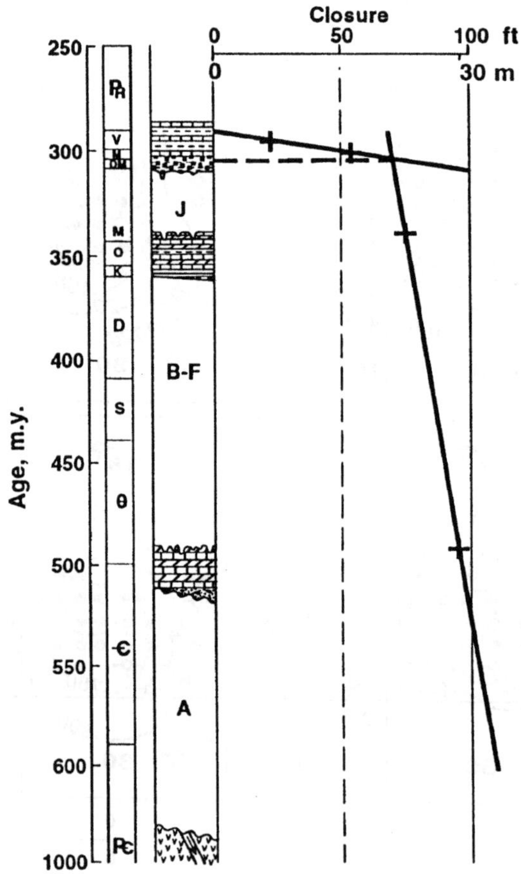

Figure 4. Crossplot showing relation of structural closure with increasing geologic age for Slick-Carson Field, Cowley County. Gaps in record noted by blanks - A, B-F, and J. (Location of field shown on Fig. 1.)

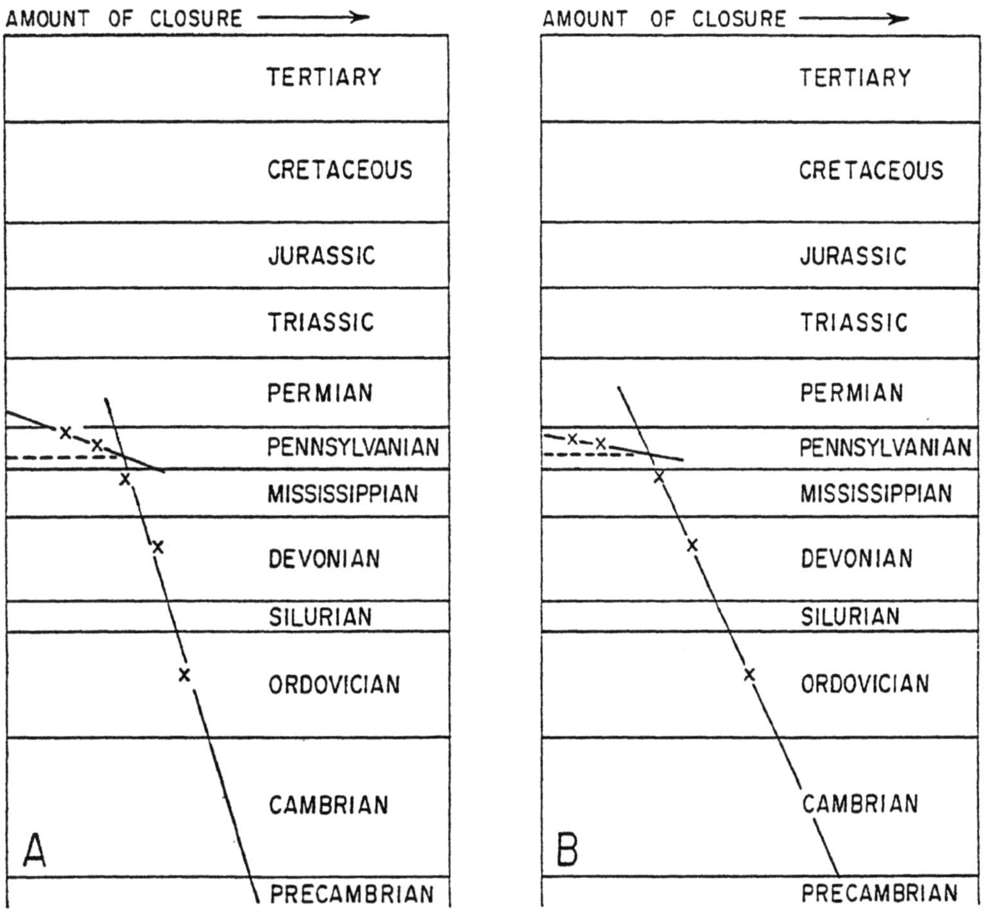

Figure 5. Crossplot showing increase of structural closure with depth for: A, Mill Creek; and B, Ashburn Fields in Wabaunsee County in Forest City Basin (from Merriam, 1963). Abrupt change in slope of line indicates time of structural deformation in Pennsylvanian. (Location of fields shown on Fig. 1.)

An alternative for the time/closure crossplot is a stratigraphic/closure crossplot. Two examples of this representation are shown in Figure 8. This crossplot highlights the unconformity at the top of the Mississippian, but is not so effective an illustration as the time/closure crossplot.

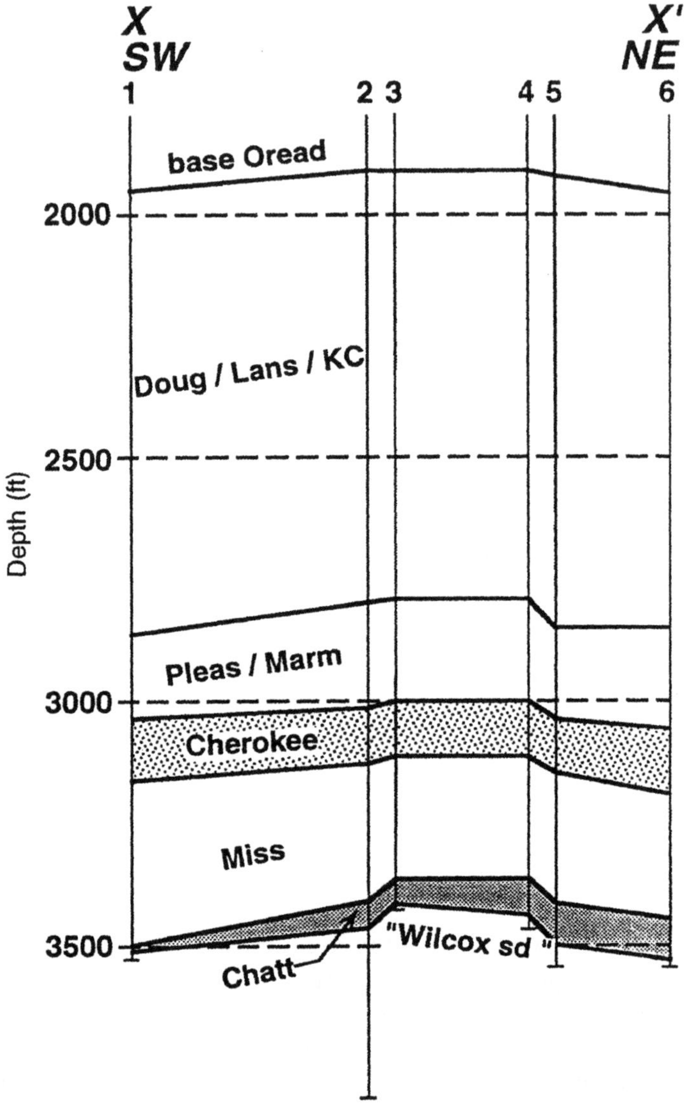

Figure 6. Cross section of Slick-Carson Field (location shown on Fig. 3). Datum sealevel. Well logs used in cross section X-X:' (1) Marland No.1 Nevitt (NE SE NW sec. 30, T32S, R3E); (2) Slick No.1 Conrod (SEc NW, sec. 19, T32S, R3E); (3) Slick No.1A Carson (SWc NE sec. 19, T32S, R3E); (4) Slick No.4B Carson (SE SE NE sec. 19, T32S, R3E); (5) Slick No.1 East Carson (SWc NW sec. 19, T32S, R3E); (6) Barnsdall No. 1 Albright (SE SW SW sec. 17, T32S, R3E).

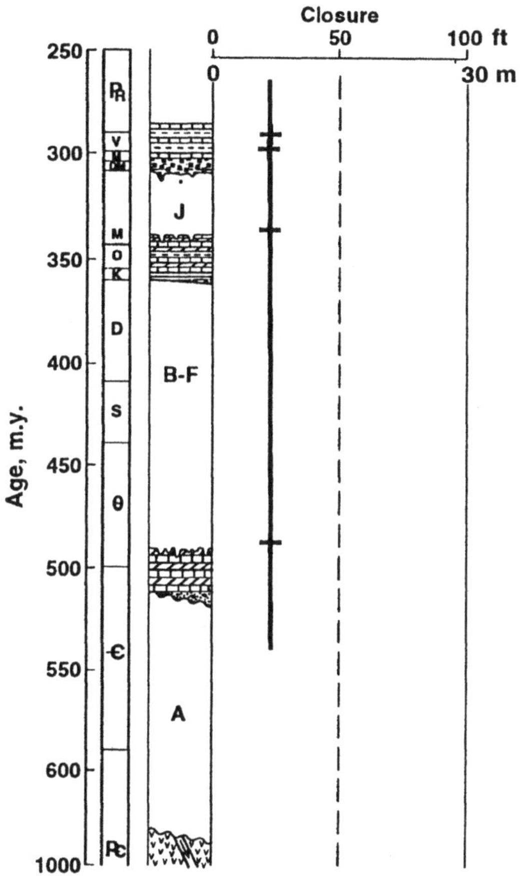

Figure 7. Crossplot showing relation of structural closure with increasing geologic age for Box Field, Cowley County. Note similar amount of closure with depth. Gaps in record noted by blanks - A, B-F, and J. (Location of field shown on Fig. 1.)

2.3. STRUCTURAL DEVELOPMENT AND RELATED ACTIVITIES

This study indicates there are two types of 'plains-type folds,' (1) those developed mainly in the early Middle Pennsylvanian as a result of the collision of the North and South American plates when the major structural features of Kansas and adjoining areas were formed; and (2) those features formed later as a result of other influences (Merriam and Förster, 1994).

The formation of the second type of features (2) is conjecture at this time without verification, which can be obtained only by geophysical investigations or by drilling. However, we would like to suggest that these features were formed by the emplacement of basic igneous intrusive plugs similar to those

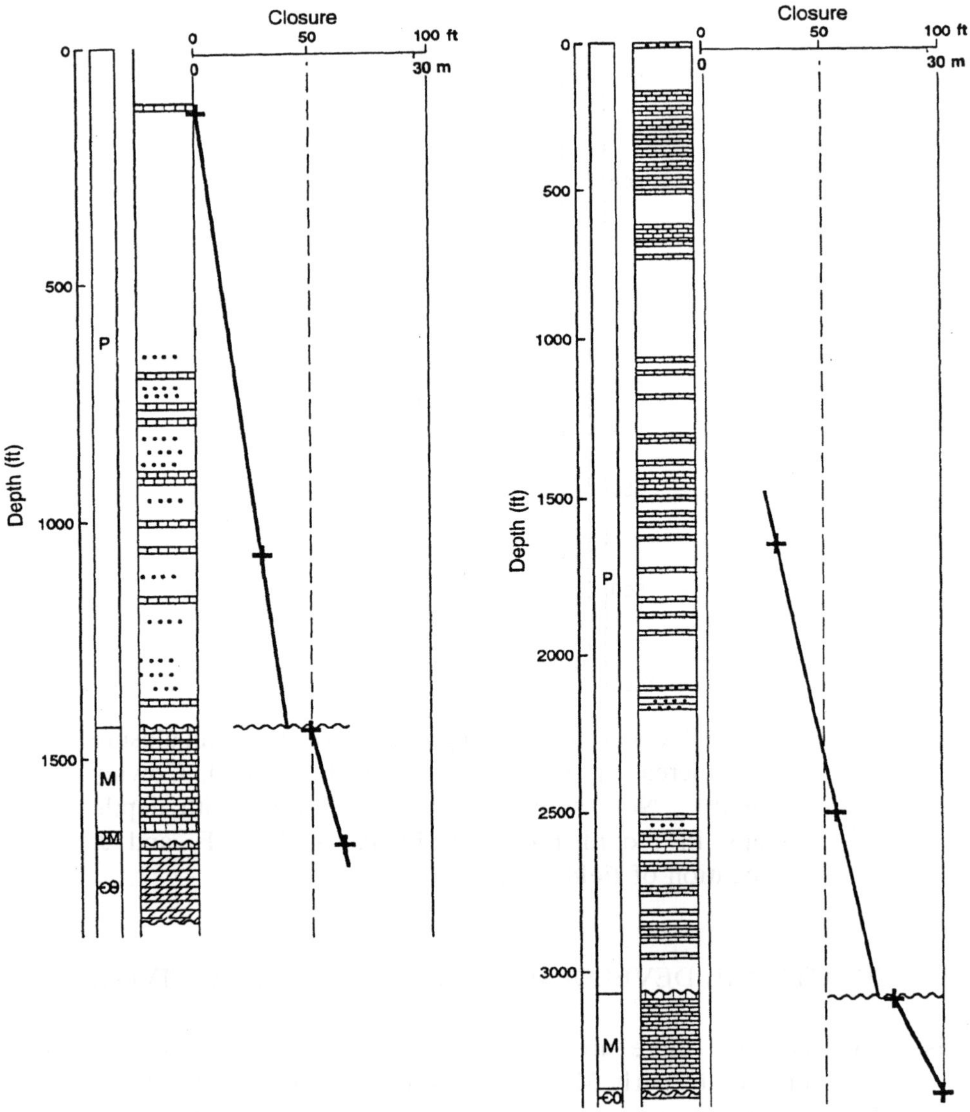

Figure 8. Crossplot of structural closure versus stratigraphic column. Diagram shows a major unconformity at top of the Mississippian/base of the Pennsylvanian.

which formed the Silver City and Rose domes in Woodson and Wilson counties. Whereas the plugs cut stratigraphic units as young as Pennsylvanian at Silver City and Rose, the plug believed to be present beneath the Box structure did not. There are several reasons for thinking Box is a feature formed differently than the other 'plains-type folds:' (1) the conformity of closure downward through the section, that is the amount of closure does not increase with depth, (2) the small and nearly circular nature of the feature, and (3) the size of the feature remains approximately the same with depth.

Other domal features in southeastern Kansas, for example the Neosho Falls Dome in northeastern Woodson County (Merriam, 1963), also are thought to be the reflection of these small intrusive features. The intrusives were emplaced in the late Cretaceous or early Tertiary. Other such intrusions are known from outcrops farther north in Riley County in the Salina Basin. The intrusions in Kansas are approximately the same age as others known in Arkansas, New York, and Wyoming. They are of economic interest because those in Arkansas contain industrial-grade diamonds and those in Kansas and Wyoming contain micro-diamonds. The thin alteration band in the sediments adjacent to the igneous material suggests that the plugs were emplaced quickly at moderate temperatures. At least some of the intrusions carry fragments of stratigraphic units penetrated by the magma. A good example of this is the granite at the Rose Dome which was brought up as fragments of Precambrian basement, which for many years was interpreted to be a granitic intrusion (Twenhofel, 1926; Knight and Landes, 1932).

2.4. TIME OF FORMATION OF 'PLAINS-TYPE FOLDS'

Once the 'plains-type fold' was formed initially, it continued to develop through time. That is chronicled by the amount of change in the thickness of stratigraphic units on and adjacent to the structures. This difference can be computed by determining the *structural gradient-depth plot*, the percent change in thickness on and off structure and plotting that change by stratigraphic unit (Merriam, 1994). Such diagrams for the four structures, Slick-Carson, Coleman, Box, and Neosho Falls Anticlines, are given in Figure 9.

Each of the structures show a slightly different time of development, which is to be expected considering that the percent change in unit thickness should show the times of rejuvenation and maximum compaction. For the Neosho Falls Dome the major time is during the Lansing-Kansas City (Fig. 9A); for Coleman during the Cherokee (Fig. 9B); for Slick-Carson during the Cherokee and Admire (Fig. 9C); and for Box during the Douglas and Admire (Fig. 9D).

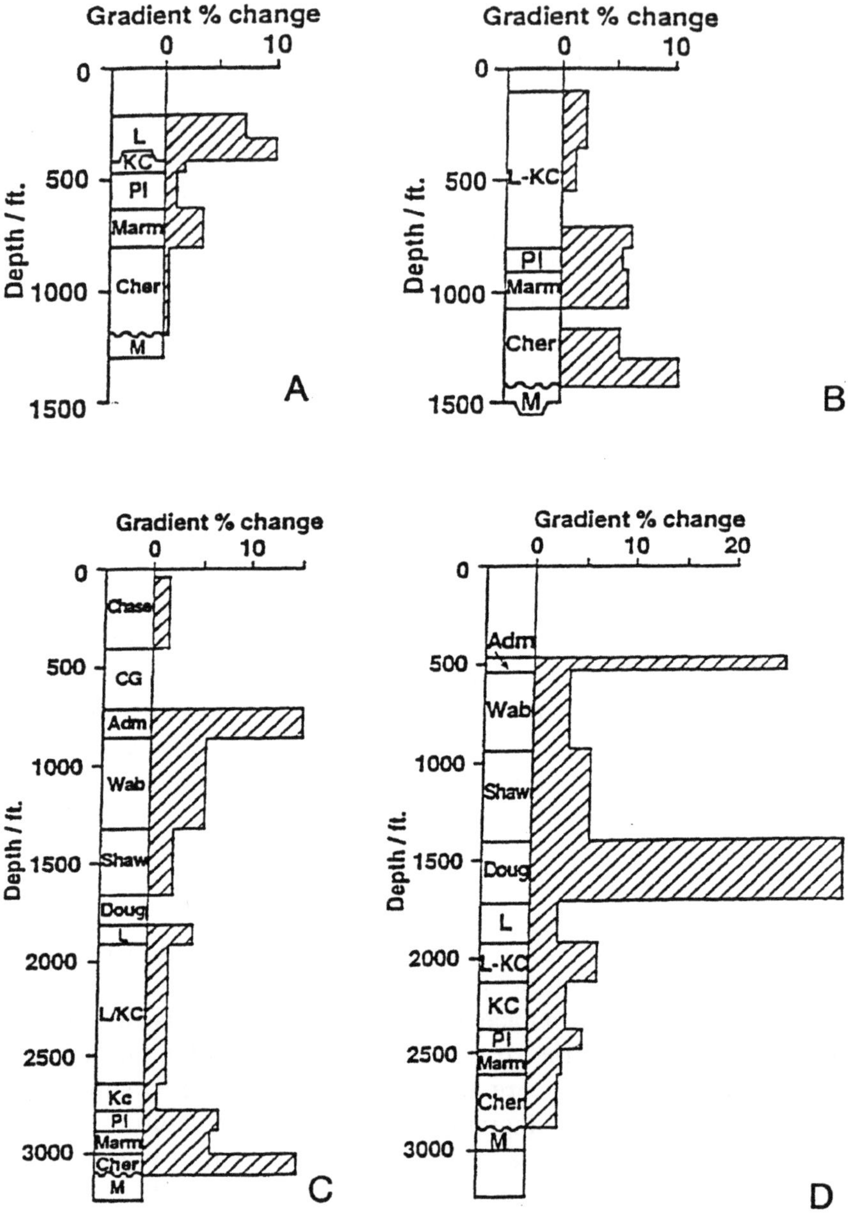

Figure 9. Structural gradient-depth plots. Percentage change in thickness on and off structures plotted versus stratigraphic unit. A high percentage of change shows the time of structural adjustment and maximum sediment compaction. A, Neosho Falls Dome; B, Coleman Field; C, Slick-Carson Field; D, Box Field.

Considering that each of the structures is on a different basement block, this is not surprising. Each of the blocks would adjust to the local stresses and move at different times depending on local conditions. By applying this technique in more detail to an area, it would be possible to determine which basement blocks moved when and if they were in place during the migration of fluids. If they were formed in time, they could have served as local traps.

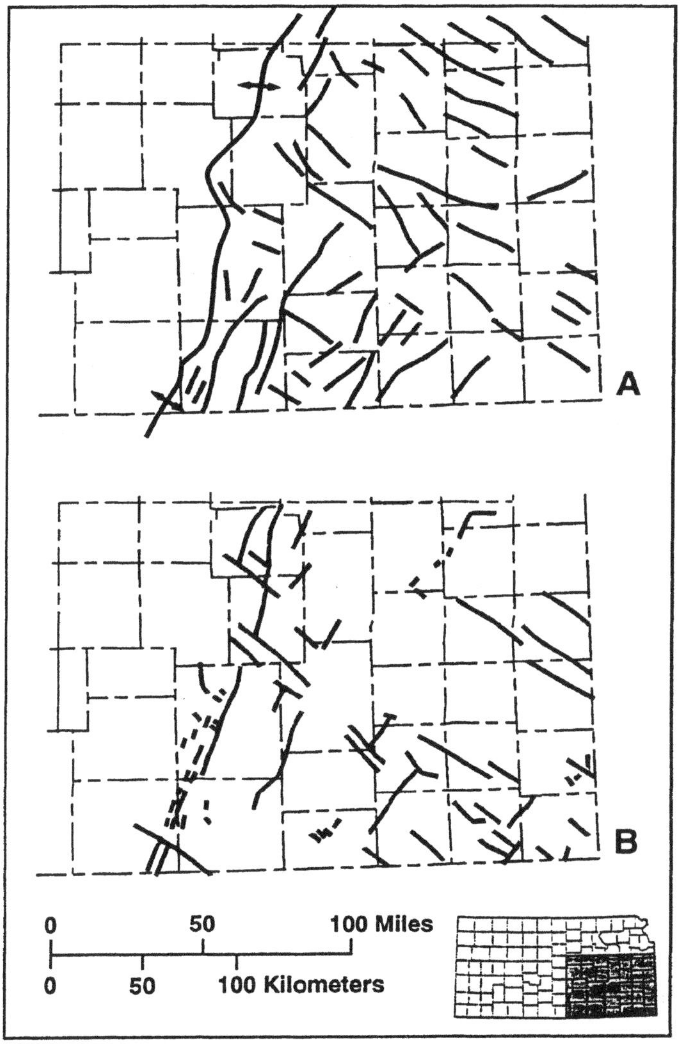

Figure 10. Direction of structural features: A, anticlinal axes in sedimentary cover; B, fractures in Precambrian basement.

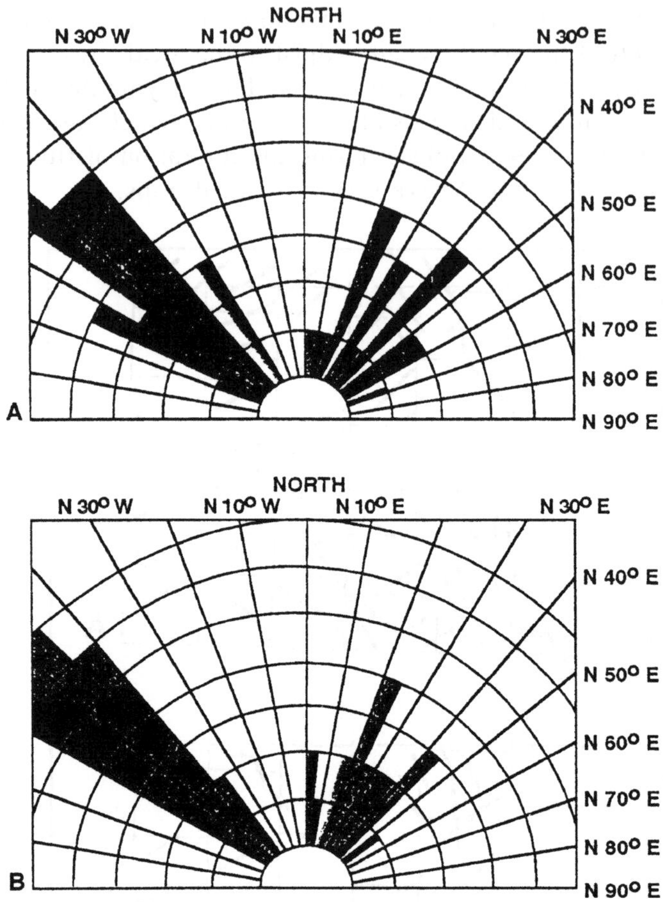

Figure 11. Rose diagrams showing direction of: A, anticlinal axes in sedimentary cover; B, fractures in Precambrian basement.

2.5. RELATION OF 'PLAINS-TYPE FOLDS' TO BASEMENT

To determine if the basement fracture pattern is reflected in the overlying sediments, the known basement fractures were mapped as well as the axes of folds in the sedimentary cover (Fig. 10). These patterns then were analyzed by noting the direction of the features and comparing the directions on rose diagrams (Fig. 11). It is obvious that the two patterns are similar and that there is little doubt that the structure in the sedimentary cover is reflecting the basement structural grain.

3. Results and Discussion

We have determined that there are two types of 'plains-type folds.' One type is the draping of sediments over rigid basement highs as a result of compaction, most probably reflecting an inherent Precambrian fracture pattern. Differential vertical structural movement of these structures is the cause for the differential compaction of the sediments. These 'plains-type folds' are small, increase in sharpness and amplitude with depth, and their crests migrate upward laterally toward the steep side of the structure.

Other 'plains-type folds' were created by the vertical emplacement of igneous plugs in the late Mesozoic or early Tertiary. These features have the same amount of closure with depth, are small and circular in nature, and the size of the structure is approximately the same with depth. Differentiation of these two types of 'plains-type folds' may be difficult without verification from other sources.

Determination of the structural gradient-depth curves can provide information on timing of the formation of structure and differential compaction and the type of 'plains-type fold.'

4. References

Bates, R.L., and Jackson, J.A., 1987, Glossary of geology (3rd ed.): Am. Geol. Inst., Alexandria, Virginia, 788 p.

Blackwelder, E., 1920, The origin of the central Kansas oil domes: Am. Assoc. Petroleum Geologists Bull., v. 4, no. 1, p. 89-94.

Clark, S.K., 1932, The mechanics of the plains-type folds of the Midcontinent area: Jour. Geology, v. 40, no. 1, p. 46-61.

Fath, A.E., 1920, The origin of the faults, anticlines, and buried 'granite ridge' of the northern part of the Midcontinent oil and gas field: U.S. Geol. Survey Prof. Paper 128C, p. 75-84.

Gardner, J.H., 1917, The vertical component in local folding: Southwestern (American) Assoc. Petroleum Geologists, v. 1, p. 107-110.

Knight, G.L., and Landes, K.K., 1932, Kansas laccoliths: Jour. Geology, v. 40, no. 1, p. 1-15.

McCoy, A.W., 1921, A short sketch of the paleogeography and historical geology of the Midcontinent oil district and its importance to petroleum geology: Am. Assoc. Petroleum Geologists Bull., v. 5, no. 5, p. 541-584.

McCoy, A.W., 1934, An interpretation of local structural development in Midcontinent areas associated with deposits of petroleum, *in* Problems of petroleum geology: Am. Assoc. Petroleum Geologists, Tulsa, p. 581-627.

Merriam, D.F., 1963, The geologic history of Kansas: Kansas Geol. Survey Bull. 162, 317 p.

Merriam, D.F., 1994, Structural gradient: a new concept in 'plains-type folding' (abst.): Kansas Acad. Sci. Ann. Meeting Abstracts (Washburn Univ.), v. 13, p. 26.

Merriam, D.F., and Förster, A., 1994, Precambrian basement control on 'plains-type folds' in the Midcontinent Region, USA (abst.): 11th Intern. Conf. Basement Tectonic '94 Abstracts (Potsdam, Germany), Intern. Basement Tectonics Assoc., p. 101-102.

Monnett, V.E., 1922, Possible origin of some of the structures of the Midcontinent oil field: Econ. Geology, v. 17, no. 3, p. 194-200.

Monnett, V.E., 1927, The 'buried hills' as a structural agency: Oklahoma Acad. Science Proc., v. 6, pt. 2, p. 268-272.

Powers, S., 1922, Reflected buried hills and their importance in petroleum geology: Econ. Geology, v. 17, no. 4, p. 233-259.

Powers, S., 1925, Structural geology of the Midcontinent region: a field for research: Geol. Soc. America Bull., v. 36, no. 2, p. 379-392.

Rich, J.L., 1935, Fault-block nature of Kansas structures suggested by elimination of regional dip: Am. Assoc. Petroleum Geologists Bull., v. 19, no. 10, p. 1540-1543.

Rubey, W.W., and Bass, N.W., 1925, The geology of Russell County, Kansas: Kansas Geol. Survey Bull. 10, pt. 1, p. 1-86.

Sugden, W., 1962, Structural analysis, and geometrical prediction for change of form with depth, of some Arabian plains-type folds: Am. Assoc. Petroleum Geologists Bull., v. 46, no. 12, p. 2213-2228.

Twenhofel, W.H., 1926, Intrusive granite of the Rose Dome, Woodson County, Kansas: Geol. Soc. America Bull., v. 37, no. 2, p. 403-412.

PART II

TITLES OF PAPERS AND POSTERS PRESENTED AT CONFERENCE

Oral contributions

Continental Scale Features of Basement Rocks of the European Craton

A. BERTHELSEN (Copenhagen, Denmark), Invited Speaker
Europe´s tectonic accretion and the intraplate origin of the Tornquist Zone

A. ZELAZNIEWICZ (Wroclaw, Poland)
Basement control over cover deformation in the Tatra MTS, West CarpathiansStrain-sensitive Mylonite as a recorder of superposed tectonic events

M. SUK (Brno, Czech Republic)
The role of Precambrian in the crust structure of central Europe

P. BANKWITZ AND E. BANKWITZ (Potsdam, Germany)
Crustal structure and development of the Saxothuringian Zone (Central Europe) from DSS data

G. ZULAUF AND W. DÖRR (Frankfurt/M., Gießen, Germany)
Cadomian tectonothermal events in the Bohemian Massif

ST. MUELLER (Zürich, Switzerland), Invited Speaker
In-situ physical properties of the earth`s crust in Central Europe

E. R. SHEIKH-ZADE (Moscow, Russia)
Comparison of the West-European and Turanian Paleozoic platforms structure

S. L KOSTIUCHENKO AND L. N. SOLODILOV (Moscow, Russia)
The accretion tectonic elements in the basement rocks of the northern part of the East European craton

Y. K. SHCHUKIN (Moscow, Russia)
Deep dynamic systems of North Eurasia

M. HEEREMANS, H. STEL, P. VAN DER BEEK AND T. PEDERSEN (Amsterdam, Netherlands; Kjeller, Norway)
Anorogenic Granites, magmatic underplating and the origin of vertical dip slip basement faults in the Fennoscandian shield

F. M. ELTER AND E. PANDELI (Siena, Florence, Italy)
Structural setting of the Palaeozoic crystalline basement of the Northern Apennines (Italy)

C. TOMEK (Brno, Tchec Republik, Salzburg, Austria)
Central European Hot Spot

G. B. FERSHTATER, F. BEA, V. N. SMIRNOV AND E. V. PUSHKAREV (Ekaterinburg, Russia; Granada, Spain)
Granodiorite compositional variations in the W-E cross section of the Urals reflecting basement tectonics and composition

Role of Basement Exhumation and Crustal Decoupling in the Evolution of Orogenic Belts

S. COSTA, P. REY, W. TODT AND ST. L. GOLDSTEIN (Mainz, Potsdam, Germany)
Consequences of post-thickening collapse on the evolution of the lower crust

J. ALCOCK AND P. MULLER (Ogontz, Oneonta, U.S.A.)
Decompression anatexis in granulite facies gneiss, northeast Adirondack Highlands, USA: Isothermal decompression before isobaric cooling

P. LEDRU (Orléans, France)
The orogenic crust of the European Variscan belt: a new light on the collision process

J. REINHARDT (Darmstadt, Germany)
Late orogenic exhumation of deeply buried rocks: Evidence for tectonic unroofing in the Saxonian Granulite Massif

O. M. ROSEN (Moscow, Russia)
Lower continental crust exhumed: Granulite grade terranes and Amphibolite grade shear zones in the Anabar Shield, Polar Siberia

V. N. PUCHKOV (Ufa Bashkortostan, Russia)
Central Uralian zone: An exhumed crystalline basement of the Laurussian continental margin

J. RÖTZLER, H. BUDZINSKI AND G. BUDZINSKI (Potsdam, Germany)
Early Variscan very high temperature (>1000°C) - high pressure metamorphism, subsequent near-isothermal uplift, and related heating of middle and upper crustal rocks in the Saxon Granulite Massif

S. N. IVANOV AND K. S. IVANOV (Ekaterinburg, Russia)
Rheological principles of the earth crust geodynamics and Russian and German super-deep drilling experiments and results

ST. MAZUR AND R. KRYZA (Wroclaw, Poland)
Thrusts and collapse in the Karkonosze-Izera Block, NE Bohemian Massif

R. HETZEL AND U. RING (Mainz, Germany)
Miocene extensional unroofing in the Menderes-Massiv, Southwestern Turkey

J. L. ALLEN (Lexington, U.S.A.)
Early Paleozoic basement uplift and subsidence on the West-Central North American craton, Colorado,

Extrapolation and Correlation of Geological Data from Basement Rocks: An Assessment

D. M. FOUNTAIN (Laramie, U.S.A.), Invited Speaker
Seismic properties of exposed lower continental crust: evidence from the lab and field

P.REY (Potsdam, Germany)
Seismic reflectors, seismic fabrics and basement tectonics

M. GRAD, A. GUTERCH, T. JANIK, R. MATERZOK, U. LUCATO, J. YLINIEMI, E. LÜCK, A. SCHULZE AND K. FÖRSTE (Warsaw, Poland; Helsinki, Finland, Potsdam, Germany)
Results of new seismic crustal refraction experiment in Permian Basin in Northwest Poland and East Germany

K. BIERBRAUER AND O. ONCKEN (Potsdam, Germany)
Structural development of deep continental crust during collision and exhumation - Implications for the seismic properties of continental crust

P. SADOWIAK, M. SIMON AND G. HIRSCHMANN (Potsdam, Hannover, Germany)
Seismic data from the crystalline basement of the KTB area (Germany) and their geological significance

M. GRAD AND U. LUOSTO (Warsaw, Poland; Helsinki, Finland)
Velocities and attenuation of seismic waves - implication for the fracturing of the uppermost basement in the Baltic Shield

J. DUYSTER, A. KONTNY AND A. RAUEN (Windischeschenbach, Germany)
The influence of microcracks and ore minerals on the electrical conductivity: An example from a fault zone in KTB

S. PARKER GAY, JR. (Salt Lake City, U.S.A.)
Mapping Precambrian plate boundaries on the cratons with residual aeromagnetic maps

A. G. GREEN (Zürich, Switzerland), Invited Speaker
Integrated geological and geophysical studies of Basement Rocks: Examples from the Canadian Lithoprobe program

M. D. THOMAS AND E. I. TANCZYK (Ottawa, Ontario, Canada)
Analysis of gravity and magnetic anomalies along Lithoprobe Trans-Hudson orogen transect, Canadian Shield

M. DIACONESCU, C. DIACONESCU, R. RADULESCU, V. RAILEANU AND A. POMPILIAN (Bucharest, Romania).
Geological and geophysical structure of the Transylvanian Basement (Romania)

H. HOSNEY, P. MORGAN, M. GHALEB AND Y. ABDELHADY (Cairo Egypt, Northern Arizona, USA)
Basement Tectonics of the Northern Gulf of Suez from geothermal and gravity data

General Themes

N. PAVONI (Zürich, Switzerland)
Late Cenozoic and recent tectonic deformation in Turkey, distributed, quasi-continuous deformation versus rigid microplate motion

U. SCHALTEGGER, F. CORFU AND J.-L.SCHNEIDER (Toronto, Canada, Zürich, Switzerland, Lille, France)
Geochronological constraints for the formation of volcano-sedimentary basins during late Variscan transtension

D. A. RODIONOV (Moscow, Russia)
Multidimensional statistical method for stratigraphical sections correlation by the complex of geological and geopysical data

I. R. STAKHOVSKY (Moscow, Russia)
Multifractal analysis of fault structures in basement rocks

A. G. GUTIÈRREZ (Salamanca, Spain)
Transition from internal to external zones, hinterland to foreland mechanisms of displacement transfer at the Narcea Antiform (NW of spain)

B. MINGRAM (Potsdam, Germany)
Lithological and geochemical composition of pressure related metamorphic units in the W-Erzgebirge, Germany

K. SCHULMANN AND P. LEDRU (Prague, Czech republic; Orléans, France)
Basement and nappes tectonics within the Bohemian Massif and French Massif Central

M. M. YOUSSEF, H. AL. IBRAHIM, A. A. BAKHEIET AND M. M. SENOSY (Cairo, Egypt)
Surface and subsurface tectonic pattern of Sohag region, middle Egypt

H. G. L. BASSI (Buenos Aires, Argentina)
Cortical Pre-Paleozoic tectonism controlling Paleozoic and Tertiary metallogeny. Central Argentina

XU GUIZHONG AND WANG YIFEN (Beiing , China)
The characteristics and tectonic environment of Precambrian basement from theTongbai Dabei orogenic belt

ST. MARSHAK AND R. S. PAULSEN (Urbana, U.S.A)
The Transamerican cratonic weak zone: A NW-trending corridor of Precambrian-Present tectonic reactivation, USA

J. HOLBROOK (Cape Girardeau, U.S.A.)
Low-amplitude welts raised throughout the Aptian/Albian U.S. Western interior as the dispersed deformational response of a shattered plate to regionally consistent intraplate stress

M. P. CARLSON (Lincoln, U.S.A.)
Effects of reactivation of the North American Midcontinent Rift system on Phanerozoic stratigraphy

M. C. GILBERT AND J. P. HOGAN (Norman, U.S.A.)
Integrated structural and igneous evolution of an ancient crustal rift

D. F. MERRIAM AND A. FÖRSTER (Lawrence, U.S.A.; Potsdam, Germany)
Precambrian basement control of "Plains-Type Folds" in the Midcontinent Region, USA

P. BERENDSEN AND P. HOTH (Lawrence, U.S.A., Potsdam, Germany)
The relationship between Proterozoic basement structure, depositional history of Paleozoic sedimentary rocks, and mineralization in the Midcontinent, U.S.A.

E. STEIN AND R. PATERSON (Darmstadt, Germany, Los Angeles,U.S.A.)
Kinematics of country rock deformation during emplacement of the Joshua Flat batholith, California, U.S.A.

D. H. ROUSSELL AND H. L. Gibson (Sudbury, Canada)
The tectonic, magmatic and mineralization history of the Sudbury structure, Canada

Poster Session

Continental Scale Features of Basement Rocks of the European Craton

U. SEBASTIAN AND ST. FACHMANN (Freiberg, Germany)
Late orogenetic basement reworking - Hercynian structural development of Cadomian granitoids in the Erzgebirge/Saxony

K. RÖTZLER (Potsdam, Germany)
An inverted and reduced metamorphic sequence of high pressure rocks superimposed on medium pressure ones in the Erzgebirge/Germany

H. G. KRÄUTNER (München, Germany)
Basement tectonics in the Romanian Carpathians - The Variscan crust south of the Alpine front

M. COLI (Firenze, Italy)
Remarks on a deep crustal section of the Northern Apennines

N. MINZONI AND M. MINZONI (Ferrara, Italy)
Geodynamic evolution of the Variscan cycle in the Sardinian-Corsican Massif

M. BERNABINI, N. BIANCHI, D. BUCCI, L. ORLANDO, M. PAROTTO AND M. TOZZI (Roma, Italy)
Basement involved tectonics and possible deep structure in recent mountain building: The central Italy case history

V. ZABIGAILO AND M. PAVLYUK (Lviv, Ukraine)
The peculiarities of basement structure of the Volyn-Podolian plate

V. K. GAVRISH AND A. A. KICHKA (Kiev, Ukraine)
Precambrian basement of the Dnieper-Donets Paleorift, Northeast Ukraine: Main features and evolution

G. O. GATTO, D. MASETTI AND N. MINZONI (Padova, Ferrara, Italy)
Pre-Mesozoic crystalline rocks in the Alpine chain of Val Venosta area (North Italy)

N. RUDITSA AND M. RAPOPORT (Ekaterinburg, Russia)
Pre-Riphean structural plan of the Urals orogenic belt (by geophysical data)

Z. SKÀCELOVÀ AND V. NEHYBKA (Brno, Czech Rep.)
Seismic activity of the Marianskè Làzne deep fault in the Western Bohemia

B. S. PANOV (Donetsk, Ukraine)
New planetary belt of deep faults in Eurasia

V. NEMEC (Praha, Czech Republic)
Regularity in the geometry of structural patterns - one of possible keys to decipher basement tectonics

M. A. BAER, G. V. KRASNOPEVTSEVA AND YU. K. SHUKIN (Moscow, Russia)
Deep conditions of formation of the European sedimentary basins

T. OBERC-DZIEDZIC, C. PIN , J. L. DUTHOU AND J. P. COUTURIE (Wroclaw, Poland; Clermont-Ferrand, France)
Petrology and Rb-Sr ages of the Strzelin granitoids, fore-Sudetic Block, SW Poland

Role of Basement Exhumation in the Evolution of Orogenic Belts

B. ALESHIN, K. ZOLOEV, V. LESTCHIKOV, B. POPOV, M. RAPOPORT, V. RYBALKA AND V. SHALAGINOV (Ekaterinburg, Russia)
The Role of the crystalline foundation of the Euroasiatic supercontinent in forming the Urals folded belt

G. CHEN (Guangzhou, P.R.C.)
Impact of basement melting on the Meso-Cenozoic tectonic development of southeast China

S. M. TILMAN AND V. D. CHEKHOVICH (Moscow, Russia)
Types of basement exhumations in orogenic belts of Northeast Russia

S. L. YUNGA AND A. F. GRACHEV (Moscow, Russia)
Joints and fractures asymmetric orientations and origin of stresses and decoupling of crustal layers at Kola Peninsula

V. I. MAKAROV (Moscow, Russia)
Neotectonic basement deformations: Types; depth of seating; mechanisms and processes

V. SOKOLOV (Sverdlovsk, Russia)
Relationships of the tectonic processes in the upper crust and the structure of the deep parts of Lithosphere

A. MARCOS AND G. GALÀN (Oviedo, Barcelona, Spain)
Structural and metamorphic evolution of high pressure basic granulites during basement exhumation: An example from the Bacariza formation, Cabo Ortegal Comlex (NW Spain)

F. J. FERNÀNDEZ AND A. MARCOS (Oviedo, Spain)
Structures developed during tectonic exhumation of high grade gneiss during Variscan orogeny (Cabo ortegal complex, NW Spain)

S. TEUFEL, L. FRANZ AND O. ONCKEN (Potsdam, Germany)
The cooling of orogenically disturbed continental crust: The Ivrea crustal segment (Northern Italy)

Extrapolation and Correlation of Geological Data from Basement Rocks: An Assessment

C. DIACONESCU, V. RAILEANU, M. DIACONESCU, F. RADULESCU, A. POMPILIAN AND M. BITER (Bucharest-Magurele, Romania)
Complex Data on the Carpathian Foredeep Basement (Romania)

N. V. FEDOROVA, F. I. NIKONOVA AND V. A. SHAPIRO (Ekaterinburg, Russia)
Regional magnetic anomaly data and other geodata of the Urals

S. KASHUBIN (Sverdlovsk, Russia)
Seismic wave anomalies and stress tectonics relationships in the Urals Crust

R. Y. STAKHOVSKAYA (Moscow, Russia)
The basement tectonics of the Caucasus Region: The model on the base of gravity data

A. RYBALKA (Sverdlovsk, Russia)
Reflectivity image of the earth´s crust near the Urals SG4 deep borehole

V. GRÜNWALD (Bergen, Norway)
Faulted basement in westernmost on-shore Norway (Oygardaen) and its offshore seismic image
V. M. MORALEV AND M. Z. GLUKHOVSKY (Moscow, Russia)
Correlation of geophysical and remote sensing data from the basement of the East-European craton and prognosis of diamond-bearing Kimberlites

M. MAMDOUH KHATTAB (Alain, United Arab. Emirates)
Magmatic activity in the Pre-Permian Basement, Southern Arabian Gulf

H. HOSNEY, M. GHALEB, S. SHARAFELDIN AND Y. ABDELHADY (Cairo, Giza, Egypt)
Gravity and magnetic interpretation of basement structure of North-Western dersert, Egypt

H. HOSNEY, Y. ABDELHADY AND S. HAKIM (Cairo, Egypt)
Ras Garra, the Gulf of Suez, Egypt: A seismo-magnetic examination of its subsurface

A. F. KAMEL (Cairo, Egypt)
Distribution of radioactivity in some granite outcrops in southeastern desert of Egypt

V. SOKOLOV (Sverdlovsk, Russia)
Changes of physical properties of the crust and seismic boundaries

General Themes

A. KH. ZILBERSHTEIN AND G. M. ROMM (St. Petersburg, Russia)
Tectonic action on integral optical characteristics of rocks

V. A. PRIVALOV AND E. A. PANOVA (Donetsk, Ukraine)
The nature of equidistant fault relationships

V. EZ (Moscow, Russia)
Detailed geological studies of metamorphic terrains and their bearing on the interpretation of the crust structure
S. V. SIROTINSKAJA (Moscow, Russia)
Integration of geological and geophysical data for prediction of ore-bearing areas

A. HENK (Würzburg, Germany)
Thermal modeling and late-orogenic exhumation - A case study

J. BERNHARD (Bergen, Norway)
Brittle deformation in the Oygarden Gneiss Complex, western Norway paleostress analysis

R.-K. ERIKSEN (Bergen, Norway)
Structure and tectonic evolution of basement in westermost onshore Norway (Oygarden Gneiss Complex)

T. O. YTREDAL (Bergen, Norway)
3D fracture pattern analysis in a basement block, Oygarden Gneiss Complex, Western Norway

P. K. VERMA (Ujjain, India)
A methodology for statistical analysis and tectonic interpretation of the remote sensing lineament fabric data

Y. Z. SUN, M. Y. LIN AND Y. J. JIN (Aachen; Hebei, China)
Coriolis force plays a part in crustal movements

A. G. GUTIÈRREZ (Salamanca, Spain)
Phyllonite development in pelites and sandstones under low-grade metamorphic conditions at the Narcea-Antiform (NW Spain)

A. NAUMENKO (Kiev, Ukraine)
Geophysical and geochemical features of the Ukrainian shield´s crust structure

M. A. FONSECA AND M. A. DARDENNE (Brazil, Brasilia)
Geometry and kinematics of basement strike - slip faults in the core of a curved fold and thrust belt: Example of the Northern Brasilia Belt, Central Brazil

Y. L. RONKIN AND S. G. CHERVJAKOVSKY (Ekaterinburg, Russia)
Isotopic geochemical evolution model of riftogenic series of acid magmatites of the Urals Western slope

P. GIESE AND D. ROEDER (Berlin, Gießen, Germany)
Review on the main features of crustal structure in Mediteranean orogenic belts - implications for deeply reaching thrusting processes

M. G. LEONOV, S. YU. KOLODYAZHNY AND M. L. SOMIN (Mocow, Russia)
Tectonic flow structures in basement and protoplatform cover

V. A. KOROTEEV AND A. I. RUSIN (Ekaterinburg, Russia)
Tectonics and metamorphism of the Urals belt basement

A. V. MASLOV AND L. A. KARSTEN (Ekaterinburg, Russia)
Riphean-Vendian complexes of the Urals as the upper Precambrian basement of the Uralides

Proceedings of the International Conferences on Basement Tectonics

1. R. Mason (ed.): *Basement Tectonics 7*. Proceedings of the Seventh International Conference on Basement Tectonics, held in Kingston, Ontario, Canada, August 1987. 1992 ISBN 0-7923-1582-0
2. J. Bartholomew, D.W. Hyndman, D.W. Mogk and R. Mason (eds.): *Basement Tectonics 8 – Characterization and Comparison of Ancient and Mesozoic Continental Margins*. Proceedings of the Eighth International Conference on Basement Tectonics, held in Butte, Montana, USA, August 1988. 1992 ISBN 0-7923-2088-3
3. M.J. Rickard, H.J. Harrington and P.R. Williams (eds.): *Basement Tectonics 9 – Australia and Other Regions*. Proceedings of the Ninth International Conference on Basement Tectonics, held in Canberra, Australia, July 1990. 1992 ISBN 0-7923-1559-6
4. R.W. Ojakangas, A.B. Dickas and J.C. Green (eds.): *Basement Tectonics 10*. Proceedings of the Tenth International Conference on Basement Tectonics, held in Duluth, Minnesota, U.S.A., August 1992. 1995 ISBN 0-7923-3342-9
5. O. Oncken and C. Janssen (eds.): *Basement Tectonics 11 – Europe and Other Regions*. Proceedings of the Eleventh International Conference on Basement Tectonics, held in Potsdam, Germany, July 1994. 1996 ISBN 0-7923-3797-2

KLUWER ACADEMIC PUBLISHERS – DORDRECHT / BOSTON / LONDON

FSC
www.fsc.org
MIX
Papier aus verantwortungsvollen Quellen
Paper from responsible sources
FSC® C105338

If you have any concerns about our products,
you can contact us on
ProductSafety@springernature.com

In case Publisher is established outside the EU,
the EU authorized representative is:
Springer Nature Customer Service Center GmbH
Europaplatz 3, 69115 Heidelberg, Germany

Printed by Libri Plureos GmbH
in Hamburg, Germany